INTERNATIONAL SERIES IN

ANALYTICAL CHEMISTRY

General Editors: R. Belcher and H. Freiser

VOLUME 59

FLUORESCENCE AND PHOSPHORESCENCE SPECTROSCOPY: PHYSICOCHEMICAL PRINCIPLES AND PRACTICE

Other Titles of Interest

BAKER and BETTERIDGE
Photoelectron Spectroscopy: Chemical and Analytical Aspects

BECKEY
Field Ionization Mass Spectrometry

DAMANY et al.
Some Aspects of Vacuum Ultraviolet Radiation Physics

ELWELL and GIDLEY
Atomic Absorption Spectroscopy

EMSLEY and LINDON
NMR Spectroscopy using Liquid Crystal Solvents

JACKMAN and STERNHELL
Applications of Nuclear Magnetic Resonance Spectroscopy
in Organic Chemistry 2nd edition

SCHEINMANN
An Introduction to Spectroscopic Methods for the Identification
of Organic Compounds (2 vols.)

SOBELMAN
An Introduction to the Theory of Atomic Spectra

FLUORESCENCE AND PHOSPHORESCENCE SPECTROSCOPY: PHYSICOCHEMICAL PRINCIPLES AND PRACTICE

by

STEPHEN G. SCHULMAN, Ph.D.

Department of Pharmaceutical Chemistry
College of Pharmacy, University of Florida

PERGAMON PRESS

OXFORD · NEW YORK · TORONTO · SYDNEY · PARIS · FRANKFURT

U.K.	Pergamon Press Ltd., Headington Hill Hall, Oxford OX3 0BW, England
U.S.A.	Pergamon Press Inc., Maxwell House, Fairview Park, Elmsford, New York 10523, U.S.A.
CANADA	Pergamon of Canada Ltd., 75 The East Mall, Toronto, Ontario, Canada
AUSTRALIA	Pergamon Press (Aust.) Pty. Ltd., 19a Boundary Street, Rushcutters Bay, N.S.W. 2011, Australia
FRANCE	Pergamon Press SARL, 24 rue des Ecoles, 75240 Paris, Cedex 05, France
WEST GERMANY	Pergamon Press GmbH, 6242 Kronberg-Taunus, Pferdstrasse 1, Frankfurt-am-Main, West Germany

First edition 1977

Library of Congress Cataloging in Publication Data

Schulman, Stephen Gregory.
Fluorescence and phosphorescence spectroscopy.

(International series in analytical chemistry; v. 59)
1. Fluorescence spectroscopy. 2. Phosphorescence spectroscopy. I. Title.
QD96.F56S38 1976   543'.085   75-46591
ISBN 0 08-020499-6

In order to make this volume available as economically and rapidly as possible the author's typescript has been reproduced in its original form. This method unfortunately has its typographical limitations but it is hoped that they in no way distract the reader.

Printed in Great Britain by A. Wheaton & Co., Exeter

To Joke and Christina
for their love, encouragement,
tolerance and infinite patience.

PREFACE

The past two decades have witnessed the emergence of fluorescence and phosphorescence spectroscopy to become among the most useful of tools in experimental biology and chemistry. The availability of low to moderate cost commercial instrumentation, in recent years, has made these analytical methods accessible to virtually all laboratories. No other instrumental methods available at comparable cost, can equal or surpass fluorimetry and phosphorimetry in analytical sensitivity. Concentrations of luminescing materials as low as 10^{-9} molar are routinely determined. This aspect is particularly desirable in the biomedical sciences where low concentrations of drugs, metabolites and toxins in blood serum and urine must routinely be monitored.

Although there are many excellent reference works currently available on luminescence spectroscopy, many of these tend to be either theoretically oriented, with heavy emphasis on quantum mechanical interpretation of spectra of simple molecules as they relate to molecular electronic energy states, or instrumentally oriented, with emphasis on electronic and optical aspects of instrument design. Most practitioners of fluorescence and phosphorescence spectroscopy, currently, are either analytical chemists or biologically oriented scientists with limited background in quantum mechanics, electronics, optics, and higher mathematics in general. This group of experimenters is usually constrained to the use of commercially available instrumentation in which there is no great deal of variability in fundamental design from one manufacturer to another. Moreover, the luminescing molecules of interest to this group are often extremely complicated drugs and metabolites which do not lend themselves to detailed understanding by rigorous quantum mechanical treatments.

Yet it is important that the limits of instrumental capability and at least a qualitative picture of the relationships
between molecular electronic structure, environmental interactions and luminescence spectra can be understood by all
practitioners in order to maximize sensitivity, selectivity,
interpretation and overall reliability of data taken in fluorescence and phosphorescence spectral measurements. This book
then, is written with the analytical chemist and biological
scientist in mind and represents an attempt to make the instrumental, and especially the structural and environmental
aspects of luminescence spectra intelligible to the reader with
a general college background in chemistry and physics.

The author expresses his gratitude to Ms. Carolyn B. Gran-
tham, Dr. A.C. Capomacchia, Dr. D.V. Naik, Mr. R.J. Sturgeon,
and Mr. Peter F. Eisenhardt for their invaluable comments and
assistance with the preparation of this manuscript.

Gainesville, Florida   Stephen G. Schulman
     July, 1976

CONTENTS

ix

CHAPTER 1

PHOTOPHYSICAL PROCESSES IN ISOLATED MOLECULES

Fluorescence and phosphorescence occur in molecules as a result of and subsequent to a series of physical phenomena, normally beginning with the absorption of light. These phenomena as well as fluorescence and phosphorescence are derived from the electromagnetic nature of light, the details of molecular structure (especially molecular electronic structure) and the nature of the environment of the luminescent molecule. It should be fairly obvious that an appreciation of these phenomena is necessary for the understanding of the relationships between molecular structure and luminescence spectroscopy to chemical and biological problems. This section then, will deal with molecular electronic structure, the interactions of light with charged particles in molecules and the events that follow the absorption of light in molecules. For the sake of simplicity, these arguments will deal only with isolated molecules. The effects of environmental interactions of molecules on electronic spectra will be considered in a later chapter.

Molecular Electronic Structure

The interactions between atomic electrons to form chemical bonds are due to the valence electrons and orbitals comprising the partially filled outer shells of atoms. The electrons occupying the filled inner shells of atoms which belong to molecules are localized upon the atoms from which they originated and contribute only very weakly, through their repulsive properties, to molecular electronic structure. Because it is the valence electrons which are responsible for the electronic spectra of molecules in the visible and ultraviolet regions of the spectrum as well as for chemical reactivity, our consideration of molecular electronic structure will be confined to

those features which arise from the valence electrons and orbitals of molecules.

A chemical bond or occupied molecular orbital may be
thought to originate from the overlap of occupied atomic orbitals. The geometry of the overlap is used to classify the
type of chemical bonding, and the filling of the molecular
orbital is governed by the Pauli exclusion principle (i.e.
a maximum of two electrons can occupy one orbital). Some of
the various types of molecular orbitals are described below.

σ-Bonds

The overlap of two atomic orbitals along the line joining
the nuclei of the bonded atoms results in a σ-bond. A σ-bond
can accommodate two electrons, in accordance with the Pauli
exclusion principle. The distribution of charge in a σ-bond
is strongly localized between the two bonded atoms. Although
each atom participating in the σ-bond contributes one atomic
orbital to the formation of the σ-orbital, the two electrons
occupying the σ-orbital may originate, one from each atom or
both from the same atom. In the former case, the σ-bond is
called a covalent bond, and in the latter case, it is called
a coordinate covalent bond.

If the two atoms joined by the covalent bond exert unequal
attractions upon the electron pair comprising the σ-bond, the
electron pair will spend more time near one atom than the other.
In this case, the more strongly attracting atom is said to be
more electronegative than the other, and the bond is said to be
a polar covalent bond. Because the electronic charge in a σ-
bond is localized along the line between two atoms, electronic
repulsion and the exclusion principle prevent the formation of
more than one σ-bond between any two atoms in a molecule.
Electrons engaged in σ-bonding are usually bound very tightly
by the molecule. Consequently, a great deal of energy is required to promote these electrons to vacant molecular orbitals.

This means that molecular electronic spectra involving σ-electron transitions occur well into the vacuum ultraviolet and are not of interest in conventional luminescence spectroscopy which is concerned with the general region between the near ultraviolet and the near infrared (i.e. 200-1000 nm).

π-Bonds

The overlap of two atomic orbitals at right angles to the line joining the nuclei of the bonded atoms is said to result in a π-bond. π-Bonding is a weaker interaction than σ-bonding and consequently, usually occurs secondarily to σ-bonding. The formation of π-bonds always involves atomic p or d orbitals, but never s orbitals. In a π-bond, the distribution of electronic charge is concentrated above and below the plane containing the σ-bond axis. While σ-electrons are strongly localized between the atoms they bind, π-electrons, not being concentrated immediately between the parent atoms, are freer to move within the molecule and are frequently distributed over several atoms. If several atoms are σ-bonded in series, and each has a p or d orbital with the proper spatial orientation to form a π-bond with the others, rather than an alternating series of localized two atom π-bonds being formed, a set of π-orbitals are formed which are spread over the entire series of atoms. These π-orbitals are said to be delocalized. In some cyclic organic molecules, π-delocalization extends over the entire molecule. These compounds are said to be aromatic and are the molecules of primary interest in fluorescence and phosphorescence spectroscopy.

Because π-electrons are not concentrated between the bonded atoms, they are not as tightly bound as σ-electrons. Hence, their electronic spectra occur at lower frequencies than do σ-electron spectra. For molecules containing isolated π-bonds, the transitions involving π-electrons are still in the vacuum ultraviolet or at the limit of the near ultraviolet. Molecules containing delocalized π-electrons usually have π-electron spectra in the near ultraviolet while the superdelocalized π-systems, the aromatic molecules, have π-electron spectra which range from the near ultraviolet for small molecules, to the near infrared for large ones.

Nonbonded Electrons

With the exceptions of most transition metal complexes and a very few stable free radicals, enough σ- and π-bonds will encompass any atom in a molecule, to engage in bonding any electrons which were unpaired in the valence shell of the isolated atom. In all atoms of the periodic table which have more than four electrons in the valence shell (e.g. nitrogen), however, there are electrons in the valence shell which are already paired. The orbitals occupied by these electrons are already filled, although the valence shell, in the isolated atom, is not complete. These electrons are unavailable for conventional covalent bonding and yet have energies comparable to other electrons in the same shell. Consequently, they are called nonbonding or n-electrons. When the atom is engaged in bonding, the valence shell electrons involved in σ-bonding will drop in energy, well below the energy of the n-electrons. The energy of the electrons involved in π-bonding (if any) will usually drop below the energy of the n-electrons but not quite as much as in the case of the σ-electrons. Because the n-electrons are higher in energy than either the σ- or π-electrons, they must be considered as potential contributors to the spectral features of molecules possessing them. While the n-electrons originate from pure atomic orbitals, the difference in repulsion experienced by these electrons as a result of the difference in electronic environment of a molecular atom compared with that of an isolated atom infer that the n-orbitals will have more of the properties of localized molecular orbitals than of atomic orbitals. This is indeed found to be the case. The bonding geometries of atoms in molecules are influenced at least as strongly by the number of nonbonded electron pairs as by the number of bonded electron pairs around them.

Coordinate Covalent Bonds

While nonbonded electron pairs in molecules do not enter into covalent bonding in the usual sense, they may be transferred into vacant molecular orbitals in suitable acceptor molecules. This results in the formation of a coordination complex in which the bond formed between the nonbonded electron pair donor and the acceptor is said to be a coordinate covalent bond. Coordinate covalent bonds are usually of the σ-type but certain molecular species also have the ability to form π-type coordination complexes. Bronsted basicity is one well-known type of coordinate covalent bond formation. In this case, the Bronsted base donated a pair of nonbonded electrons to a vacant 1s orbital of a hydrogen ion to form the conjugate acid. The σ-bond formed between the base and the hydrogen ion results in the loss of identity of the nonbonded pair on the base. The formation of coordination complexes has significance in the interpretation of spectra of compounds having nonbonded electron pairs. This subject will be discussed at length in a later section.

Antibonding Orbitals

In addition to the low energy molecular orbitals which comprise the chemical bonds of molecules, each molecule has associated with it a series of higher energy molecular orbitals which are unoccupied under ordinary conditions. These orbitals represent regions of space to which electrons may be promoted if sufficient energy is imparted to the molecule and are called antibonding molecular orbitals. Antibonding orbitals may have σ-symmetry (i.e. the electron density lies along the bonding axis) and are then denoted as σ*-orbitals or they may have π-symmetry (i.e. the electron density lies out of the plane containing the bonding axis) and are then denoted as π*-orbitals.

Electronic absorption entails the promotion of an electron, by the absorption of energy, from an originally occupied bonding or nonbonded orbital (σ, π or n) to an originally unoccupied

molecular orbital. Fluorescence and phosphorescence entail
the demotion of an electron from a σ*- or π*-orbital to a σ-,
π- or n-orbital, with the release of energy.

In all organic molecules except free radicals, the bonding
and nonbonding orbitals are normally doubly occupied and the
electrons in these orbitals have opposite spin quantum numbers,
in accordance with the Pauli exclusion principle. However,
when an electron is promoted to a higher orbital the orbital
originally occupied and the antibonding orbital occupied are
both singly occupied. The Pauli exclusion principle then no
longer requires that the two electrons have opposite spins and
there is at least some probability that one electron will change
spin. The distinction between fluorescence and phosphorescence
will later be seen to depend upon whether or not the promoted
electron changes spin.

Orbitals and States

The electrons occupying molecular orbitals are each characterized by certain physical properties, namely orbital angular
momentum, energy spin angular momentum. The orbital angular
momentum results from the path the electron follows in the
molecule which is in turn derived from the orbital symmetry
(σ or π). The energy of an electron in a molecule is determined
by the electrostatic attractions experienced by the electron
from the atomic nuclei in the molecule and the repulsions from
other molecular electrons. These attractions and repulsions
are dependent upon the orbital occupied by the electron. Electrons occupying σ-bonds which are constructed from sp^2 atomic
hybrid orbitals experience greater average nuclear attractions
than those occupying orbitals constructed from sp^3 hybrids.
This is a result of the greater fraction of s character in the
sp^2 hybrid and the fact that s orbitals have higher electron
density at the nuclei than p orbitals. The spin angular momentum is thought to originate from the electron spinning on its
own axis.

The way in which the electrons of a molecule are distributed amongst the molecular orbitals is called the electronic configuration of the molecule. Each electronic configuration of a molecule has a resultant orbital angular momentum, spin angular momentum and energy which is comprised of the vector sums of the orbital angular momenta and spin angular momenta, respectively, and the scalar sum of the energies of each electron belonging to the molecule. The angular momenta and electronic energy of the molecule are said to specify the electronic state of the molecule. In most organic molecules of spectroscopic interest it is only the π-electrons and n-electrons which are involved in spectroscopic phenomena and in order to simplify the treatment of molecular electronic structure, all other electrons (i.e. the σ-electrons and the inner shell electrons localized on individual atoms) are generally neglected. Spectroscopic transitions are incapable of determining the absolute energies of electronic states but rather determine the differences in energy between states. The zero of energy of a molecule is thus arbitrarily taken to be the lowest lying electronic state. This is generally referred to as the ground electronic state of the molecule and corresponds to the state having the configuration in which all electrons are in the lowest energy orbitals available. Promotion of an electron from an orbital which is occupied in the ground state to one which is normally unoccupied in the ground state is said to result in an electronically excited state of the molecule. Because there are several unoccupied orbitals for each molecule, several electronically excited states are possible for each molecule. Each electronic state of a molecule is characterized by a particular distribution of electronic charge in the molecule. This means that the dipole moment of a molecule in an electronically excited state will generally be different from that of the ground state. The distinction between orbital energies and configurational or state energies is represented graphically in Fig. 1.1.

The electronic states of a molecule are classified

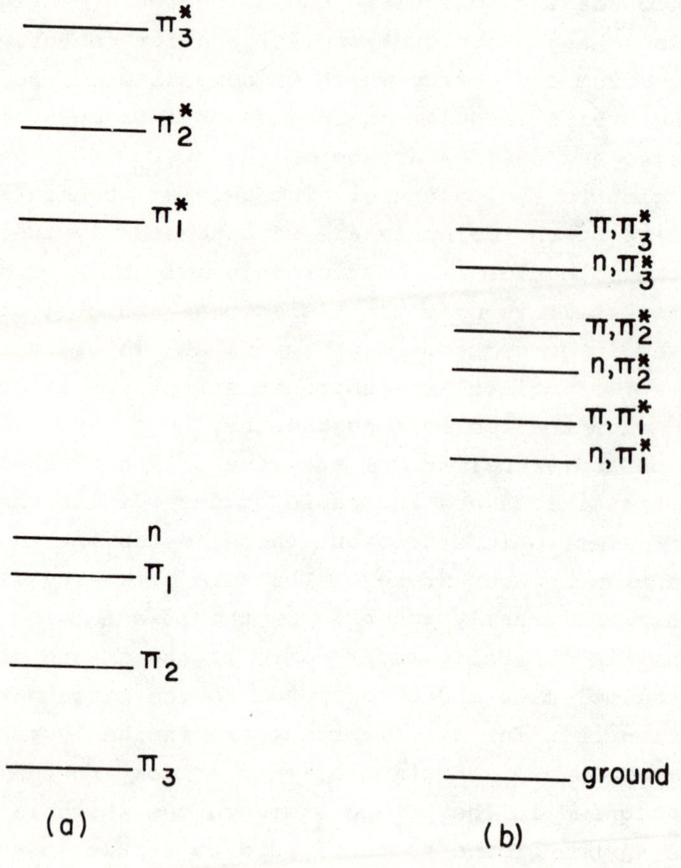

$$
\begin{array}{ll}
\underline{\hspace{2cm}} & \pi_3^* \\[1.5em]
\underline{\hspace{2cm}} & \pi_2^* \\[1.5em]
\underline{\hspace{2cm}} & \pi_1^* \\
\end{array}
$$

$$
\begin{array}{ll}
\underline{\hspace{2cm}} & \pi,\pi_3^* \\
\underline{\hspace{2cm}} & n,\pi_3^* \\[0.8em]
\underline{\hspace{2cm}} & \pi,\pi_2^* \\
\underline{\hspace{2cm}} & n,\pi_2^* \\[0.8em]
\underline{\hspace{2cm}} & \pi,\pi_1^* \\
\underline{\hspace{2cm}} & n,\pi_1^* \\
\end{array}
$$

$$
\begin{array}{ll}
\underline{\hspace{2cm}} & n \\
\underline{\hspace{2cm}} & \pi_1 \\[1.5em]
\underline{\hspace{2cm}} & \pi_2 \\[2em]
\underline{\hspace{2cm}} & \pi_3 \\
\end{array}
$$

$$\underline{\hspace{2cm}} \quad \text{ground}$$

$$(a) \qquad\qquad\qquad (b)$$

Fig. 1.1. (a) The molecular orbitals of pyridine in order
of increasing energy.

(b) The lowest electronic states of pyridine in order
of increasing energy. In the ground state π_1, π_2,
π_3 and n in (a) are each occupied by two electrons. The electronically excited states (n, π_1^*,
π, π_1^*, etc.) are produced by promoting an electron from n or π, to π_1^*, π_2^* or π_3^*. The n,π^*
states higher than n,π_1^* are not observed in the
absorption spectra because they are weak and
buried under the more intense π,π^* transitions.

(labelled) according to the spin and orbital angular momenta associated with each state. There are several classification systems currently in use in electronic spectroscopy and of these two, the Schonflies point group notation and the Platt free electron model are by far the most widely used. The system used to label electronic states in this text will be that derived from the Platt free electron model.

The Platt free electron model of molecular electronic structure is an approximate quantum mechanical model of symmetrical cyclic aromatic hydrocarbons which treats π-electrons as if they were free to move about the perimeter of the hydrocarbon molecule. Each electron and each state of the molecule is described by a wave-like periodic probability function which has maxima, minima and nodes much like a classical wave. The ground electronic state of any molecule in the Platt system is denoted by the letter A. The lower excited states (those which are observed spectroscopically) are denoted by L_a, L_b, B_a, B_b, C_a and C_b. The excited states denoted by the letter L correspond to states of high orbital angular momentum and are lower in energy than the B or C states. Those designated by the letters B and C correspond to states of low orbital angular momentum. The subscripts a and b refer to the nodal properties of the states as derived from the formalism of the free electron model. States having the subscript a have nodes passing through the centers of the bonds and maximum electron density at the atoms of the molecule. States having the subscript b have nodes passing through the atoms and have maximum electron density at the bond centers. The ground state A has no nodes and therefore has no subscript designation.

Electronic states are also designated by spin angular momentum. In most organic molecules the ground state has an even number of electrons which are paired in the lowest lying molecular orbitals. There are two possible orientations of the spin angular momentum vector of an electron and in the ground states of most organic molecules the number of electrons with

either spin vector orientation is equal, hence the net spin
angular momentum is zero. States which have zero spin angular
momentum (i.e. no unpaired electrons) are called singlet states
because if an external magnetic field is placed across the
molecule having no unpaired electrons the spin angular momentum
vector can have only one discernable orientation with respect to
the magnetic field. The complete designation of the ground
state of a molecule with no unpaired electrons is thus given by
^1A, which is read as "singlet A".

If an electron is promoted from an occupied orbital to a
higher, previously unoccupied orbital the electrons promoted
will usually retain its original spin in which case the excited
state will also be a singlet state and may be designated ^1L_a
(read "singlet L_a"), ^1L_b, ^1B_a, ^1B_b, ^1C_a or ^1C_b. If, however,
the promoted electron does change spin, there will be two un-
paired electrons in the molecule (the promoted electron and the
residual electron). In a molecule with two unpaired electrons
there are three possible orientations of the resultant spin
angular momentum vector in an externally applied magnetic
field (in the same direction as the field vector, opposite to
the field vector and normal to the field vector). Thus, a
molecule with two unpaired electrons is said to be in a triplet
excited state. The state designations then become ^3L_a (read
"triplet L_a"), ^3L_b, ^3B_a, ^3B_b, ^3C_a and ^3C_b. In general, the
number of orientations the spin angular momentum vector can
have in an external magnetic field is called the multiplicity
of the state and is equal to the number of unpaired electrons
plus one.

The Platt system of classification of electron states just
described, applies only to aromatic molecules having π-elec-
trons and no nonbonding (n) electrons. The ^1L_a, ^1L_b, ^1B_a, ^1B_b,
^1C_a and ^1C_b designations for excited states thus correspond
to states formed by the promotion of a π-electron to an unoccupied (π*) orbital and are in general termed π,π* excited
states. The singlet excited states are collectively designated

as $^1\pi,\pi*$ and the triplet states $^3\pi,\pi*$. In these molecules the L_a and L_b states are always the lowest lying and are thought to be the states from which fluorescence and phosphorescence occur, almost exclusively. It is not possible to tell a <u>priori</u> whether the ^1L_a or ^1L_b state will be the lowest excited singlet state or whether the ^3L_a or ^3L_b state will be the lowest triplet state. However, it is found experimentally that in the smaller aromatics (i.e. those derived from benzene and naphthalene) the ^1L_b state is lowest of the singlet states while in higher aromatics the ^1L_a state lies lowest. There is also experimental evidence that the ^3L_a state is almost invariably the lowest triplet state. The distinction between the ^1L_a and ^1L_b states is frequently made on the basis of substituent effects, because of the generally greater sensitivity of the ^1L_a state in benzenoid molecules to substituents and the dependences of the energies of the ^1L_a and ^1L_b states to the positions of substituents in polycyclic aromatic molecules. This will be discussed in greater detail when substituent effects on absorption spectra are considered in the next section.

The triplet state arising from any given electronic configuration generally lies lower in energy than the singlet state of the same configuration, in isolated molecules, because the promoted electron and the residual electron in the triplet state are, on the average, farther apart, resulting in lower repulsion (destabilization) in the triplet state. Thus the ^3L_a state lies below the ^1L_a state and the ^3L_b state lies below the ^1L_b state.

In molecules having nonbonded electron pairs, it is possible to have excited states formed by the promotion of an n-electron to a vacant π-orbital. Excited states formed in this way are called n,$\pi*$ states and singlet (^{1}n,$\pi*$) states and triplet (^{3}n,$\pi*$) states of this type are both possible. While the bonding π-orbitals are much lower in energy than the atomic orbitals from which they are formed, as a result of the stabilization derived from the bonding interaction, the n-electrons

are only slightly lower in energy than atomic valence shell
electrons. As a result, the n-electrons are higher in energy
than π-electrons in the same molecule and the energy gap between the n-orbitals and π*- (vacant π) orbitals is lower than
the gap between the π- and π*-orbitals. Hence, n,π* excited
states generally lie lower than π,π* excited states involving
the same π*-orbital. In aromatic molecules, n-orbitals are
most often encountered in heteroatoms in the aromatic ring and
in carbonyl or thiocarbonyl groups. The n-orbital is usually
an sp^2 hybrid directed in the plane of the aromatic ring or at
right angles to the π- and π*-orbitals. Because the π- and π*-
orbitals are parallel or collinear, the promoted and residual
electrons in π,π* excited states are closer, on the average,
than in n,π* excited states. As a result, there is less repulsion between electrons in n,π* excited states and the lowest
^{3}n,π* state does not lie as far below the corresponding ^{1}n,π*
excited state as does the lowest ^{3}π,π* excited state below its
corresponding singlet state. The significance of this is that
while the lowest ^{1}n,π* state almost always lies below the lowest
^{1}π,π* state, the lowest ^{1}n,π* state frequently lies below the
lowest ^{3}n,π* state. This will be seen to have significance in
the interpretation of phosphorescence phenomena.

Not all nonbonded electron pairs lie perpendicular to the
π- and π*-orbitals. Some, such as those in exocyclic amino
groups and hydroxyl groups are almost parallel to the π-orbitals of the aromatic ring and in this case may actually partic-
ipate in the π-electron structure of the aromatic system. The
electron donating properties of hydroxyl and amino groups in
electrophilic aromatic reactivity are well-known and are examples of participation of nonbonded electron pairs in the π-
system. The excited states formed by the promotion of these
electrons (which will hereafter be called lone-pair or l-
electrons to distinguish them from n-electrons) are generally
found to behave more like π,π* states than n,π* states. These
π- and π*-like states will hereafter be called intramolecular
charge transfer states (or l,π* states) because they arise from
the transfer of an electron from the exocyclic group to the

aromatic ring.   In n,π* states the promoted electron is more
or less localized near the group from which it originated but
in 1,π* states the promoted electron may be displaced far from
its site of origin.   Metal to ligand electron transfer in
coordination complexes results in a special case of the 1,π*
state.

Intramolecular charge transfer excited states may also
arise from the promotion of a π-electron from the aromatic ring
to a vacant π*-orbital localized on an exocyclic group.   This
phenomenon is often observed in aromatic carboxylic acids,
aldehydes and ketones where the carbonyl type group is conjugated with the aromatic ring and in certain transition metal
complexes where π-electrons from the aromatic ligand can be
transferred into vacant d orbitals on the metal ion.   Although,
strictly speaking, the classification of states derived from
the Platt free electron model is not applicable to intramolecular charge transfer states, the labels derived from the Platt
model can be applied loosely, for the sake of convenience, to
these states, at least for the comparison of substituent effects
upon electronic spectra.   The relationship of charge transfer
donor and acceptor orbitals to the π-structure of aromatic
molecules is shown in Fig. 1.1.

In addition to the electronic states already described,
it is possible to have excited states arising from the promotion of σ-bond electrons to π*-orbitals (i.e. σ,π* states)
and from the promotion of σ-, π- or n-electrons to σ*-orbitals
(σ,σ*,π,σ* and n,σ* states).   However, the σ-orbitals lie so
low in energy and the σ*-orbitals so high in energy that the
spectroscopic phenomena involving these orbitals usually lie
in the vacuum ultraviolet region of the electromagnetic spectrum (< 200 nm) which is out of the useful range of conventional
absorption, fluorescence and phosphorescence spectrophotometers.
Therefore, these states and spectroscopic phenomena involving
them will not be considered further in this text.   Moreover,
the well-known ligand field transitions arising from the d

orbitals of transition metal ions which lie in the visible and
infrared regions of the spectrum are so weak as to be of little
analytical utility at the present time. Hence, the ligand field
states of transition metal ions and the electronic transitions
involving them will not be treated here.

The Vibrational Substructure of Electronic States

The arrangement in space of the atoms comprising a molecule
determine the molecular geometry. The atoms, however, are not
rigidly fixed in space but execute periodic motions with respect
to one another and with respect to the center of mass of the
molecule. These periodic motions of the molecular atoms are
called normal vibrations. They result from the tendency of the
positively charged nuclei to repel each other and the tendency
of the bonding electrons to hold them together. Because of the
mobility of the bonding electrons, the nuclei never come to rest
but rather, vibrate about an equilibrium position. The close
interrelationship between electronic and vibrational structure
arises from the dependence of both the electronic and vibrational properties on the electronic distribution (state) of the
molecule. The valence electrons may be thought of as the
"springs" which hold the vibrating molecule together. These
"springs" will have restorative properties which vary with the
electronic distribution in the molecule. As a result, the
equilibrium positions of the component nuclei (geometry) of
the molecule may be different for different electronic states
of the same molecule.

Just as it is possible to produce electronically excited
states by the alteration of the electron distribution of the
molecule, it is possible to produce vibrationally excited
states by distorting the molecular geometry. These states
are vibrationally excited by comparison with the lowest energy
vibrational mode possible for the molecule (the energy of
which is called the zeroth vibrational level) and their attainment requires the absorption of energy in steps (or quantum

jumps) each of which is of the order of ten percent of the
magnitude of the energy required to promote an electron to a
higher orbital. Because of differences in electronic distributions, each electronic state of a molecule has its own group
of associated vibrational energy levels (vibrational manifold)
and to a fair approximation, the total energy of a molecule is
given by the sum of its electronic and vibrational energy.

In our future consideration of the features of electronic
spectral bands, which is the reason for belaboring here the details of vibrational electronic (vibronic) structure, it will be
assumed, for the sake of simplicity, that in any electronic
state of any molecule at thermal equilibrium with its environment, only the lowest (zeroth) vibrational level of that electronic state is populated. In other words, all molecules are
assumed to be in the lowest possible vibrational state, a
situation which is actually achieved only at absolute zero.

We shall now proceed to a consideration of the intercon-
versions between molecular electronic and vibrational states
that contribute to molecular electronic spectra.

The Absorption of Light by Molecules

The Interaction of Light with Molecular Electronic Structure

A light wave may be thought of as an electromagnetic dis-
turbance traveling in a straight line with an in vacuo speed
(c) of 3.0×10^{10} cm/sec. At right angles to the direction of
travel of the wave there is an alternating electric field and
at right angles to the direction of travel of the wave and to
the plane of oscillation of the electric field vector there is
an alternating magnetic field. The frequency of oscillation of
the electric and magnetic field vectors (the number of times
per second that the field vector is at a maximum) is said to be
the frequency of the light (ν). The distance traveled by the
wave during the period of one complete oscillation is called

the wavelength of the light (λ). The speed, frequency and wavelength of the light wave are related by the equation:

$$C = \lambda\nu \qquad (1.1)$$

Because of the electric field associated with light, a charged particle (e.g. an electron) placed in the path of a light wave will experience a force and is capable of absorbing energy from the electric field of the light wave. If an electron belonging to a molecule in its ground electronic state absorbs energy from the electric field of a light wave, the electron will be promoted to an unoccupied orbital and will be transported from one site to another in the molecule. The net result will be that the molecule will have absorbed energy from the light and will be raised from the ground state to an electronically excited state. However, not all frequencies of light are capable of being absorbed by molecular electrons. Quantum theory tells us that the energy associated with one wavelength of light of frequency ν is

$$E = h\nu = h\frac{c}{\lambda} \qquad (1.2)$$

where h is a proportionality constant known as Planck's constant (h = 6.625 x 10^{-27} erg sec). A necessary condition for light of frequency ν to be absorbed by a molecule in its ground state is that the energy gap between the ground state and excited state to which excitation occurs is exactly equal to $h\nu$, or

$$E_e - E_g = h\nu \qquad (1.3)$$

where E_g and E_e are the energies of the ground and excited states, respectively. If $E_e - E_g$ is not equal to $h\nu$, absorption will not occur and the molecule is said to be transparent to light of frequency ν. The absorption of light by a molecule is said to be a type of electronic transition called electronic absorption or electronic excitation. Because the distribution of molecular electronic charge (i.e. the dipole moment) changes when light is absorbed, the absorption of light by a molecule is also called an electronic dipole transition. The dipole moments associated with the various electronically excited states of a molecule are generally different and the line along which the resultant dipole moment changes on going from the ground state to any given excited state is called the direction of polarization of the electronic transition.

The electronic transition in which we are interested here (the n → π*, π → π* and intramolecular charge transfer transitions of aromatic molecules) are all affected by light of 200-1000 nm in wavelength (the near ultraviolet and visible regions of the electromagnetic spectrum).

Electronic Absorption Spectra

The electronic absorption spectrum of a molecule is a graphical representation of the intensity of light absorbed in producing electronic transitions in the molecule, as a function of the frequency (or wavelength) of the light. In regions where the intensity of light absorbed is high are said to occur strong absorption bands. In regions of frequency where the intensity of light absorbed is low, weak absorption bands are said to occur. Although most absorption spectra are represented as absorbance (absorption intensity) vs wavelength, a proper absorption spectrum represents absorbance vs frequency because the frequency is linearly related to the energy gaps between the states involved in the electronic transitions which form the bases of the spectral bands while the wavelength is hyperbolically related to frequency and energy. The reason that most spectra are recorded linear in wavelength rather than frequency is because the dispersion elements (monochromators) of spectral instruments often disperse light linearly as a function of wavelength, and it is thus easier to calibrate monochromators with constant scanning speed in terms of wavelength.

In the smaller aromatic molecules (e.g. those derived from
benzene and naphthalene) having no n-electrons, three bands
corresponding to $\pi \rightarrow \pi*$ transitions are normally observed in
the visible and near ultraviolet regions of the spectrum.
These bands correspond to transitions from the ground state to
the ^1L_a, ^1L_b and ^1B_b states (i.e. the $^1A \rightarrow {}^1L_a$, $^1A \rightarrow {}^1L_b$ and
$^1A \rightarrow {}^1B_b$ transitions, respectively). In larger aromatic mole-
cules (e.g. those derived from anthracene, tetracene, etc.)
the bands corresponding to the $^1A \rightarrow {}^1B_a$, $^1A \rightarrow {}^1C_a$ and $^1A \rightarrow$
^1C_b transitions may also be observed and the entire $\pi \rightarrow \pi*$
absorption spectrum goes to lower frequency with increasing
size of the aromatic system. Transitions from the ground state
to the excited triplet states are highly improbable because
the change in spin accompanying light absorption constitutes
a violation of the law of conservation of spin angular momentum.
Hence, the transitions to the triplet states do not appear to a
measurable extent in the absorption spectrum. The absorption
bands corresponding to the $^1A \rightarrow {}^1L_a$, $^1A \rightarrow {}^1L_b$, $^1A \rightarrow {}^1B_a$, $^1A \rightarrow$
^1B_b, $^1A \rightarrow {}^1C_a$ and $^1A \rightarrow {}^1C_b$ transitions are called the ^1L_a, ^1L_b,
^1B_a, ^1B_b, ^1C_a and ^1C_b absorption bands, respectively. The ^1L_a
and ^1L_b states are the lowest lying $^1\pi,\pi*$ states and thus the
^1L_a and ^1L_b bands are the lowest frequency or highest wavelength $\pi \rightarrow \pi*$ bands occurring in the absorption spectrum of any
aromatic molecule. In fluorescing molecules fluorescence occurs
almost exclusively from the ^1L_a or ^1L_b state and further
discussion of $\pi \rightarrow \pi*$ absorption spectra will be restricted to
the ^1L_a and ^1L_b bands.

The relative positions of the ^1L_a and ^1L_b bands in the
absorption spectra of aromatic molecules cannot be predicted
a priori. However, in small aromatics (e.g. those derived
from benzene and naphthalene) the ^1L_b band is usually at lowest
frequency and in higher linearly annelated aromatics (derived
from anthracene, tetracene, etc.) the ^1L_a band is at lowest
frequency. Intramolecular charge transfer bands usually
occur at much lower frequencies than ^1L_a and ^1L_b bands in the
parent hydrocarbon. The ^1L_b band can often be distinguished

from the ^1L_a band because it is usually weaker in intensity and
more structured. In aromatic molecules derived from benzene,
the position of the ^1L_a band is usually more sensitive to substituents than is that of the ^1L_b band because of the greater
charge densities at the atoms of the molecule in the ^1L_a state.
Substituents in the aromatic ring perturb the charge distributions in the molecule thereby altering the energy of the ^1L_a
state and the position of the ^1L_a band appreciably. In aromatic
molecules derived from linearly annelated hydrocarbons, the ^1L_a
transition is polarized along the short axis of the molecule
and the ^1L_b transition along the long axis of the molecule (Fig.
1.2). Hence, substituents in α-type positions affect most the
position of the ^1L_a band while substituents in β-type positions
affect most the position of the ^1L_b band. Experimentally, band
assignments are usually made in molecules by preparing derivatives substituted in various positions in the molecule and
noting which absorption bands are most affected by substitution
in a given position. Occasionally, the ^1L_a and ^1L_b bands overlap to such an extent that identification of the maximum of
either band is impossible. Preparation of substituted derivatives often results in separation of the two bands and makes
identification possible.

Aromatic molecules having n-electrons on atoms participat-
ing in the aromatic structure may show $^1n \rightarrow \pi*$ bands at lower
frequencies than any of the $^1\pi \rightarrow \pi*$ bands. This is a result of
the n-orbitals in the ground state of the molecule being higher
in energy than the occupied π-orbitals. n- and π*-bands are
usually ten to one hundred times weaker than the ^1L_a or ^1L_b
bands because the n-orbitals are directed at right angles to
the π*-orbitals while the π- and π*-orbitals are parallel or
collinear. This results in poorer orbital overlap between the
n- and π*-orbitals than between the π- and π*-orbitals and
makes for lower transition efficiency in n → π* promotion. Due
to their low intensities, n → π* transitions are often not
directly observed even though they are present because they
may be buried in the low frequency side of the lowest frequency

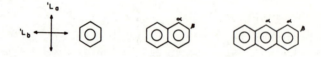

Fig. 1.2.   Illustrating the directions of polarization of the
            two lowest singlet $\pi \to \pi^*$ transitions ($^1L_a \leftarrow {}^1A$ and
            $^1L_b \leftarrow {}^1A$) in benzene, naphthalene, anthracene and
            linearly annelated polyacenes, in general. Substit-
            in α-type positions have the greatest effect on the
            $^1L_a \leftarrow {}^1A$ transition. Substituents in β-type positions
            have the greatest effect on the $^1L_b \leftarrow {}^1A$ -transition.

π- or π*-band. In these cases, the evidence for the existence
of n- and π*-bands is usually indirect, coming from the interference of the low lying $^1n,\pi*$ state with the fluorescence
properties of the molecule. This will be discussed in detail
in the next section.

The theoretical treatment of the intensities of absorption
bands is beyond the scope of this text. However, experimentally
it is found that the intensity of light absorbed in producing
a given transition is given by the Lambert-Beer law

$$\frac{I}{I_o} = 10^{-\varepsilon C \ell} \qquad\qquad (1.4)$$

in which I and I_o are the intensities of light transmitted
through the absorber and incident upon the absorber, respectively and ε, C and ℓ are, respectively, an intensity factor
characteristic of the transition ion in the particular molecule
called the molar absorptivity, the concentration of absorbing
molecules in the sample and the length of the path of light
through the sample. I/I_o represents the fraction of light
waves of the proper frequency to produce the electronic transition, absorbed on passage through the sample. For chemical
purposes, Eqn. (1.4) is usually represented in logarithmic
form

$$A = -\log \frac{I}{I_o} = \varepsilon C \ell$$

in which A is called the absorbance or optical density of the
sample. The experimentally determined intensity of transition
in a sample containing only one absorbing species, A is
linearly related to the conjugation of absorber and the optical
thickness of the sample, both of which are independent of the
nature of the transition, and the molar absorptivity which is
a function of the nature of the transition. For any given absorbing species, ε is independent of absorber concentration or
sample size and depends only on the transition properties (i.e.
absorption frequency and transition efficiency which are derived

from orbital properties). Hence, the proper comparison of
absorptive transition intensities is the comparison of molar
absorptivities for the various transitions. An electronic
absorptive transition actually encompasses the entire absorption
band and to be accurate the comparison of molar absorptivities
should be made on the basis of the integrated band intensity
(i.e. $\int \varepsilon(\nu)d\nu$). However, the evaluation of the integrated band
intensity is cumbersome and often impossible in the case of
overlapping absorption bands. Consequently, for Gaussian bands
ε is usually evaluated at the band maximum (a single frequency)
because of the approximate proportionality between the maximum
height and the area of a Gaussian band. This treatment has its
limitations, however, especially in the case of overlapping
bands where the distortion due to overlap may displace the
apparent maximum from the true maximum and when weak bands
appearing as shoulders on more intense bands may make the weak
transition appear much more intense than it really is.

For purposes of comparison, ^1L_a bands normally have molar
absorptivities from 1000-10,000, ^1L_b bands from 250-10,000 and
$^1n \rightarrow \pi^*$ bands from 10-1000.

Molecular electronic absorption spectral bands are not
thin lines occurring at one frequency as Eqn. (1.3) might imply.
Rather, they are spread over several thousands of wavenumbers
(wavenumbers are expedient frequency units equivalent to reciprocal wavelength in cm) and may appear featureless and bell-
shaped or may demonstrate clear or at least discernable fine
structure (i.e. several closely spaced peaks may appear in one
spectral band). The breadth and fine structure of molecular
electronic absorption bands are results of the dependence of
atomic motions (vibrations) in the molecule upon the changes
in charge distributions produced by electronic transition.

In order to develop an approximate theory of the way in
which molecular vibrations affect electronic absorption bands,
it is necessary to consider the effect that electronic charge

distribution in ground and electronically excited states has
on the vibrations of absorbing molecules. This treatment would
be extremely difficult if it were not for the fact that in most
molecules electronic reorganization accompanying transition
occurs much faster ($\sim 10^{-15}$ sec) than nuclear motions (> 10^{-14}
sec). The latter represents a statement of the Franck-Condon
principle. The Franck-Condon principle simplifies the treatment
of thermal relaxation processes (vibration) by allowing these
processes to be treated after the fact of electronic transition,
rather than at the same time. Consider a collection of molecules initially in the ground electronic state. The ground
electronic state is characterized by a particular electronic
distribution which is the most stable for that molecule.
Associated with the electronic and nuclear configuration of the
ground electronic state (and with those of all higher electronic
states) there are several vibrational states, each characterized
by a particular oscillatory frequency and equilibrium nuclear
configuration, and each slightly different from the others in
energy. It will be assumed that initially all of the molecules
are in the lowest vibrational level of the ground state (i.e.
the collection of molecules is ideally thermalized). Now let
us assume that the collection of molecules is made to absorb
light of such wavelength that transition occurs to an electronically excited singlet state of the molecular species. Furthermore, let us assume that the exciting light is of sufficient
band width that absorption occurs to all vibrational sublevels
of the electronically excited singlet state (Fig. 1.3). The
vibrational sublevels of each electronic state will be denoted
by the vibrational quantum numbers v = 0,1,2,3... (in order of
increasing energy). As a result of the vibrational substructure of the electronic states involved in the absorption
process, the absorption spectrum corresponding to the ground →
excited singlet transition will not consist of a single line,
as in atomic spectra, but will be a band of closely spaced
vibrational-electronic (hereafter called vibronic) lines
corresponding to the 0-0, 0-1, 0-2, 0-3 ... subtransitions.
The breadths of the vibronic subbands are due to rotational

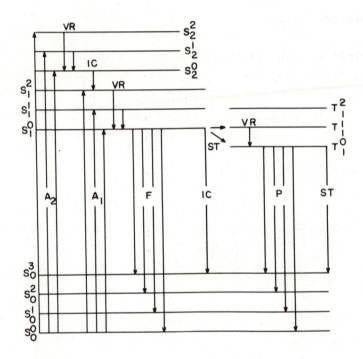

Fig. 1.3. Photophysical events in molecules. A_1 and A_2
          represent electronic absorption resulting in ex-
          citation from the ground state S_0 to the excited
          singlet states S_1 and S_2. Absorption is followed
          by vibrational relaxation (VR) or radiationless
          internal conversion (IC) between electronic states.
          The superscripts of the electronic states represent
          the vibrational quantum numbers of the various
          vibrational sublevels. Return to the ground state
          from S_1^0 can proceed by internal conversion, fluo-
          rescence (F) or intersystem crossing (ST) to the
          lowest triplet state (T_1). The triplet state is
          deactivated by phosphorescence (P) or radiationless
          intersystem crossing.

substructure while the relative intensities of the vibronic
subbands are related to the quantum mechanical efficiencies of
the various vibronic bands. These subjects are beyond the
scope of this work and for further information the reader is
referred to the excellent references in the bibliography at the
end of this chapter.

In rigid aromatic ring molecules and in certain aromatics
containing functional groups, vibrational substructure is
generally quite pronounced in electronic absorption spectra.
However, in certain aromatics having functional groups which
rehybridize and accept or donate substantial electronic charge
to the aromatic ring upon absorption, the vibrational fine
structure of absorption spectra may be lost or blurred. This is
a result of loss of vibrational quantization in the excited
state because of rapid nuclear adjustment accompanying intramolecular charge transfer resulting from excitation. The loss
of vibrational structure in the intramolecular charge transfer
band may then be said to be the result of failure of the ap-
plicability of the Franck-Condon principle to the case in
question.

Vibrational Relaxation and Internal Conversion:

Thermal Equilibration of Electronically Excited Molecules

The absorption of light by a molecule leaves the molecule
in one of a number of possible vibrational levels of one of its
electronically excited states. Although the absorption process
is extremely rapid, taking about 10^{-15} sec, the sequence of
events that return the excited molecule to its ground state is
considerably slower, taking from 10^{-14} to several sec. Return
to the ground state via molecular luminescence is among the
slowest of processes in electronically excited states, requiring
from 10^{-9} to a few sec. However, thermal equilibration by loss of
vibrational energy is much faster requiring from $10^{-14} - 10^{-12}$

sec. Consequently, fluorescence and phosphorescence always
originate from thermally equilibrated, electronically excited
molecules (we shall assume that thermal equilibration corresponds to all molecules being in the lowest vibrational level
($v = 0$). In this section, the events in the lifetimes of electronically excited states of molecules which yield the thermally
equilibrated, luminescing species will be discussed. A schematic
representation of these events is depicted in Fig. 1.3.

Vibrational Relaxation

Upon arriving in the electronically excited state, the
excited molecule may be in a vibrationally excited state (with
$v > 0$). The molecule will then begin to vibrate with a frequency characteristic of the vibrationally excited state, giving
up stepwise, its excess vibrational energy in the form of infrared quanta or in the form of kinetic energy imparted to other
molecules with which it collides. Within the lifetime of a few
vibrations ($10^{-14} - 10^{-12}$ sec) it will have thermally relaxed to
the lowest vibrational level of the electronically excited
singlet state (i.e. the electronically excited molecule attains
thermal equilibrium with the environment).

Internal Conversion

Once an excited molecule has relaxed to the lowest vibrational level of the electronically excited state, it can lose
excitation energy only by going to a lower electronic energy
level. This can be accomplished in one of three ways. If the
higher vibrational levels of the lower electronic state overlap
the lower vibrational levels of the higher electronic state
(that is if the nuclear configurations and energies of the two
electronic states are identical during a low energy vibration
of the upper electronic state and a high energy vibration of
the lower electronic state) the upper and lower electronic
states will be in a transient thermal equilibrium which will
permit population of the lower electronic state. Vibrational
relaxation of the lower electronic state then follows as before.

Crossover from a higher to a lower excited singlet state by this vibrational coupling mechanism is called internal conversion.

If the lower vibrational levels of the upper electronic state do not overlap the higher vibrational levels of the lower electronic state but are separated by a small gap (a few vibrational quanta wide), the molecule in the upper electronic state may still convert to the lower electronic state by a tunnelling mechanism. Tunnelling, the transmission of particles through regions which are classically forbidden to the particles, is a well-known phenomenon in quantum mechanics and has been employed to explain radioactivity and solid state phenomena as well as spectroscopic phenomena. This subject is discussed at length in many excellent texts on quantum theory. The probability of tunnelling decreases as the difference in energy between the lower vibrational levels of the upper electronic state and the upper vibrational levels of the lower electronic state increases.

If the energy separations of the upper and lower electronic states are relatively great, so that direct vibrational coupling is impossible and tunnelling is improbable, a third process competes with the other two for direct depopulation of the upper electronic state. The third process consists of a radiative transition from the lowest vibrational level of the upper electronic state to any one of a number of vibrational levels of the lower electronic state. The excess energy in this case is released as a wave of visible or ultraviolet light whose frequency depends upon the difference in energy between the lowest vibrational level of the upper electronic state and the vibrational level of the lower electronic state to which the radiative transition occurs. Following radiative transition, the molecule undergoes vibrational relaxation to the lowest vibrational level of the lower electronic state. The radiative deactivation of the upper electronically excited state is known as fluorescence.

Whether an excited molecule employs internal conversion

or fluorescence to pass from a higher electronic state to a
lower one depends upon the difference in energy between the
upper and lower states and the number of vibrational states
associated with each electronic state. The latter property is
important because it determines the probability of vibrational
overlap of the electronic states. If a molecule has a large
number of modes of vibration in the lower state, then the lower
vibrational modes of the higher electronic state will be able
to excite higher vibrational modes of the lower electronic
state with high probability. Hence, vibrational dissipation of
excitation energy and thus internal conversion will be favored.
It is for this reason that aliphatic molecules and others which
do not have rigid molecular skeletons and thus have many vibrational degrees of freedom, rarely exhibit fluorescence. The
aromatic molecules, with their rigid ring structures have many
fewer vibrational degrees of freedom than aliphatic molecules
and are the class of compounds which most often show fluorescence.

In most molecules, the excited electronic states are much
closer together than are the ground state and the lowest excited
state. This is seen in the overlap of absorption spectral bands
in most aromatic molecules. Consequently, efficient internal
conversion almost always precludes the possibility of fluorescence arising as the result of electronic transition between
states other than the ground state and the lowest excited state
of the same multiplicity as the ground state. Exceptions to
this are azulene and some of its derivatives in which the lowest
frequency and second absorption bands are widely separated and
in which fluorescence takes place from the second excited
singlet state to the ground state.

The internal conversion process is very rapid, taking
about 10^{-12} sec. The average lifetime of the thermally equilibrated, lowest excited singlet state, however, is of the order
of 10^{-8} sec. Consequently, even if a molecule cannot pass
efficiently from its lowest excited singlet state to the ground

state by internal conversion, it may undergo other processes during the lifetime of the lowest excited singlet state which may compete with fluorescence.

Deactivation of the Thermally Equilibrated

Lowest Excited Singlet State

Upon arriving in the lowest vibrational level of the lowest excited singlet state an electronically excited, isolated molecule still has a tendency to give up energy in order to attain the state of lowest potential energy, the lowest vibrational level of the ground electronic state. If the energy gap between the ground and lowest excited singlet state is small, the excited molecule may undergo internal conversion to the ground state giving up its excess energy in molecular vibrations. Under ordinary conditions of excitation (low light intensity) this process is not physically observable. However, if the energy gap between the lowest excited singlet state and the ground state is appreciable, the lowest excited singlet state may be depopulated by fluorescence or by radiationless intersystem crossing to the lowest triplet state at the expense of fluorescence.

Singlet-Triplet Intersystem Crossing

Because there is less electronic repulsion in any given triplet state than in the singlet state of the same electronic configuration, the triplet state lies below the excited singlet state in energy. Usually there is substantial overlap between the lower vibrational levels of the excited singlet state and the upper vibrational levels of the triplet state and there is some probability that the triplet state can be populated from the excited singlet state by a mechanism akin to internal conversion. Because the population of the triplet state from the singlet state involves a change in spin angular momentum, the process is classically forbidden (i.e. has zero probability).

However, quantum mechanically there is a finite probability
that change of spin can occur but the probability is much lower
than for the corresponding process with no change of spin.   In
terms of rate processes, the reciprocal of the probability per
event is the time taken per event or lifetime of the process in
question. The result of the much lower probability (10^6 fold)
of spin-forbidden vibrational processes occurring as opposed to
spin-allowed processes is that the mean lifetime of the spin-
forbidden process is much longer than the corresponding spin-
allowed process. Spin-allowed vibrational transitions (vibrational relaxation and internal conversion) have mean lifetimes
of $\sim 10^{-14}$ sec. Therefore, the mean lifetime of the spin-forbidden vibrational transition (intersystem crossing) is about
10^{-8} sec, which is about the mean lifetime of a fluorescing
molecule. Consequently, while intersystem crossing is too slow
to compete with internal conversion, it is of the proper time
scale to compete with fluorescence for deactivation of the lowest excited singlet state. Moreover, while most aromatic molecules undergo some degree of intersystem crossing from the lowest excited singlet state, those containing atoms of high atomic
number, n-electrons or transition metal ions employ intersystem
crossing as an efficient, if not exclusive means of deactivating
the lowest excited singlet state. This is the result of a
phenomenon called spin-orbital coupling, which consists of
addition of the spin and orbital angular momenta, vectorially,
so that the spin angular momentum of the molecule, per se, is
not well defined. This partially removes the distinctiveness
and the forbiddenness of singlet triplet transitions. In molecules containing transition metal ions or heavy atoms (e.g.
iodine) and in those having an $^1n,\pi*$ state as the lowest excited
singlet state, intersystem crossing is so efficient that these
types of molecules are seldom fluorescent. If the lowest excited singlet state is of the $^1\pi,\pi*$ type, rapid thermal relaxation will bring the excited molecule to the lowest vibrational
level of the lowest triplet state. However, if the lowest excited singlet state is of the $^1n,\pi*$ type, the lowest triplet
state may be of the $^3n,\pi*$ type as in benzaldehyde and

acetophenone, but will more often be of the $^3\pi,\pi^*$ type. This
is a result of the greater separation of the $^1\pi,\pi^*$ and $^3\pi,\pi^*$
states than that of the $^1n,\pi^*$ and $^3n,\pi^*$ states (Fig. 1.4) caused
by the greater difference in repulsion between n,π^* states and
π,π^* configuration. This difference in repulsion between n,π^*
states and π,π^* states often results in the $^3\pi,\pi^*$ state lying
below the $^3n,\pi^*$ state, even though the $^1n,\pi^*$ state lies below
the lowest $^1\pi,\pi^*$ state. If the lowest excited singlet state is
of the $^1n,\pi^*$ type, and the lowest triplet of the $^3\pi,\pi^*$ type,
intersystem crossing populates first the $^3n,\pi^*$ state which then
undergoes rapid vibrational relaxation and internal conversion
to the $^3\pi,\pi^*$ state. The carbonyl compounds derived from the
larger aromatic rings, several of the benzaldehydes and aceto-
phenones substituted with electron donating groups, and the N-
heterocyclics all have lowest triplet states which are of the
π,π^* type. The lowest triplet state may be depopulated by
several alternative pathways which will be discussed later.

Fluorescence

As an alternative to intersystem crossing, the molecule
may remain in the lowest vibrational level of the lowest excited
singlet state for $\sim 10^{-8}$ sec and then emit visible or ultraviolet fluorescence, depending upon the energy gap between the
excited state and the ground state. The fluorescing molecule
will then arrive in any one of several vibrational levels of
the ground electronic state. Although the lifetimes of fluorescent molecules are typically of the order of 10^{-8} sec, the
actual electronic transition (fluorescence) occurs within the
time scale typical of electronic transitions ($\sim 10^{-5}$ sec), so
that there is insufficient time for vibrational relaxation to
occur during the transition. Thus the fluorescing or phosphorescing molecule is likely to arrive in a vibrationally excited level of the ground electronic state even though emission
originates from the lowest vibrational level of the lowest
excited singlet state. Subsequent to emission, thermal relaxation occurs in $\sim 10^{-12}$ sec, with the molecule ultimately
arriving in the lowest vibrational level of the ground state.

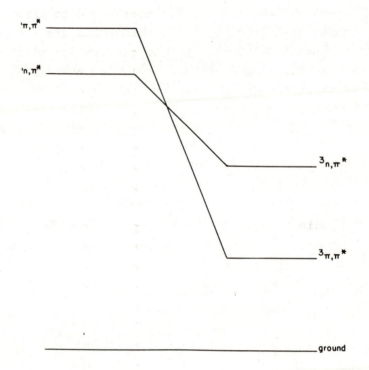

Fig. 1.4. Illustrating the inversion of n,π* and π,π* triplet
          states relative to the order of the singlet states
          derived from the corresponding electronic configura-
          tions; a result of less interelectronic repulsion
          in the ^{3}π,π* state than in the ^{3}n,π* state due to
          greater charge separation in the more delocalized
          ^{3}π,π* state.

Because vibrational relaxation processes in the excited state occur after light absorption, they are not reflected in the electronic absorption spectra. Similarly, because vibrational relaxation processes in the ground state occur subsequent to light emission, they are not reflected in the fluorescence spectra. Since all absorptions are considered to arise from the same vibrational level of the ground state and terminate in the various vibrational levels of the excited state (the unrelaxed or Franck-Condon excited state), the absorption spectra reflect the vibrational structure of the Franck-Condon excited state. Because all fluorescence arises from the same vibrational level of the lowest excited singlet state and terminates in the various vibrational levels of the ground state (the Franck-Condon ground state), the fluorescence spectrum reflects the vibrational structure of the Franck-Condon ground state.

The vibronic transitions in absorption and fluorescence spectra which correspond to the transitions between the lowest vibrational level of the ground state and the lowest vibrational level of the excited state (the 0-0 bands) are identical in energy, and correspond to the long wavelength vibronic feature of the longest wavelength absorption band and the short wavelength vibronic feature of the fluorescence band. If the Franck-Condon ground and lowest excited singlet states have the same vibrational structures (i.e. the vibrational spacings are the same in both), the fluorescence and longest wavelength absorption bands will be mirror images of each other, with the absorption spectrum lying on the short wavelength side and the fluorescence spectrum lying on the long wavelength side of the 0-0 band (Fig. 1.5). However, owing to the differences in the charge distributions in the ground and excited states, the vibrational structures of these states are often somewhat different, so that at best an imperfect mirror image relationship frequently results.

Although the electronic absorption spectrum of a molecule may contain several absorption bands corresponding to several

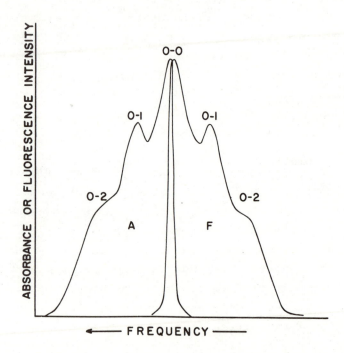

Fig. 1.5. Illustrating the mirror image relationship between
          fluorescence (F) and the lowest frequency absorption
          band (A) of a molecule in the gas phase or in a non-
          polar solvent, and in which the spacing between
          vibrational sublevels is identical in ground and
          lowest electronically excited states. 0-0, 0-1, and
          0-2 represent the vibronic transitions between the
          lowest vibrational level of the initial state and
          the various vibrational levels of the terminal state
          in each transition.

transitions from the ground state to the various excited singlet
states, the fluorescence spectrum will invariably show only one
emission band, even if a higher frequency absorption is excited,
because internal conversion and vibrational relaxation in the
higher excited states is so rapid that emission cannot compete
with radiationless deactivation of the upper states. Consequently, fluorescence only occurs from the lowest excited
singlet state to the ground state. The presence of more than
one fluorescence band in the fluorescence spectrum of a molecule
is indicative that fluorescence is occurring from more than one
chemical species. Most often, impurities account for extraneous
fluorescence bands.

Because fluorescence originates from the lowest excited
singlet state and terminates in the ground state, the effects
upon the position of the fluorescence band of substituents and
size of the aromatic system (i.e. extension of conjugation) are
qualitatively similar to those upon the position of the lowest
frequency absorption band. Hence, increasing size of the aromatic system results in lower frequencies of fluorescence because the energy gap between the ground and excited states is
compressed as the degree of conjugation in the aromatic molecule
increases. As examples of this, benzene, naphthalene, anthracene and tetracene, which have 1, 2, 3 and 4 linearly annelated
rings, respectively, fluoresce ultraviolet ($\bar{v}_f = 3.82 \times 10^4$
cm^{-1}), ultraviolet ($\bar{v}_f = 3.18 \times 10^{-4}$ cm^{-1}), blue ($\bar{v}_f = 2.64 \times$
10^4 cm^{-1}), and green ($\bar{v}_f = 2.08 \times 10^4$ cm^{-1}, respectively. Aromatic rings containing single substituents such as $-OH, -NH_2$
(excited state electron donors) or $-\overset{O}{\overset{\|}{C}}-$ (excited state charge
acceptors or vacant orbital donors) are more extensively con-
jugated in their lowest excited singlet states than in their
ground states. The stabilization of the excited state by exocyclic conjugation, relative to the ground state, results in
longer wavelengths of absorption and fluorescence relative to
the unsubstituted aromatic compounds. Substituents which do
not conjugate appreciably with the aromatic ring (e.g. $-NH_3^+$,
$-CH_3^+$, $-SO_3^-$, affect the excited state to a much lesser degree

than strongly conjugating substituents and accordingly produce considerably smaller spectral shifts. In singly substituted benzenes and naphthalenes, the lowest excited singlet state is invariably the ^1L_b state (in nonpolar solvents). Because substituents in β-positions do, the fluorescence maxima of β-substituted naphthalenes are generally more sensitive to substituents than those of α-substituted naphthalenes. In anthracene and higher linear polyacenes, the ^1L_a state is lowest and α-substitution generally produces larger spectral shifts than β-substitution. When an aromatic ring contains more than one substituent, the shift of the fluorescence spectrum relative to the parent hydrocarbon tends to approximate the vector sum of the shifting effects of each substituent. In 0- and m-disubstituted benzenes, the ^1L_b band is most affected, while in p-disubstituted benzene the ^1L_a band is displaced most from that of benzene. Since fluorescence in substituted benzenes originates from the ^1L_b state, the 0- and m-disubstituted derivatives emit farthest away from the fluorescence of benzene. In the linear polyacenes, the geometrical aspects of several interacting substituents become far more complicated and will not be considered further here.

Because other processes (e.g. intersystem crossing and internal conversion) compete with fluorescence for deactivation of the lowest excited singlet state, of a given number of molecules in the lowest vibrational level of their lowest excited singlet state at any instant of time, not all will return to the ground state by fluorescence. The fraction of excited molecules that do fluoresce is called the quantum yield of fluorescence (ϕ_f), or fluorescence efficiency, and under given conditions of temperature and environment is a physical constant of the excited molecular species. In terms of the rates of processes competing for deactivation of the lowest excited singlet state, ϕ_f is given by

$$\phi_f = \frac{k_f}{k_f + \Sigma k_d} \tag{1.5}$$

where k_f is the molecular probability that the excited molecule
will fluoresce (rate constant for fluorescence), and Σk_d is the
sum of the rate constants for deactivation of the lowest excited
singlet state by all competitive radiationless processes (for
isolated molecules, Σk_d is the sum of the rate constants for
internal conversion and intersystem crossing). The greater the
number and the greater the rates of radiationless processes,
the smaller will be ϕ_f. ϕ_f usually decreases with increasing
temperature. This is a result of the population of higher
vibrational levels of the lowest excited singlet state, by
increased temperature, which contributes to the increase in Σk_d
and the deactivation of the excited state by vibrational (nonradiative) pathways. The reciprocal of k_f is given by $\tau_f^{\,o}$ and
is called the radiative lifetime of the lowest excited singlet
state. This is the mean time the excited molecule would spend
in the excited state if fluorescence was the only means of deactivation of the excited state. The reciprocal of $k_f + \Sigma k_d$
is called the lifetime of the lowest excited singlet state (τ_f)
and corresponds to the actual mean time the molecule spends in
the excited state. The quantum yield of fluorescence may be
expressed in terms of $\tau_f^{\,o}$ and τ_f

$$\phi_f = \frac{\tau_f}{\tau_f^{\,o}} \tag{1.6}$$

and since ϕ_f is a fraction (never greater than 1), $\tau_f^{\,o}$ is
always greater than or equal to τ_f. In general, the greater
the number of processes competing with fluorescence for deactivation of the lowest excited singlet state and the greater
their rate constants the shorter will be the actual lifetime of
the lowest excited singlet state. Molecules having heavy atoms
or nonbonded electron pairs usually have high rates of intersystem crossing. Moreover, molecules having an $^1n,\pi^*$ state as
the lowest excited singlet state have long radiative lifetimes
as a result of the poor overlap between n- and π^*-orbitals.
As a result, intersystem crossing is usually 100 percent effective in deactivating the lowest excited singlet states of molecules having lowest $^1n,\pi^*$ excited states, and fluorescence is

very rarely observed from these molecules.

The relationship between fluorescence intensity and molecular properties related to the transition can be derived from
the Lambert-Beer law (Eqn. 1.4). If $I/I_o = 10^{-\epsilon C \ell}$ is the
fraction of light intensity transmitted in exciting a transition
whose molar absorptivity is ϵ at the wavelength of excitation
in a sample whose molar concentration is C, and has an optical
path length ℓ, then the fraction of light absorbed is

$$\frac{I_a}{I_o} = 1 - \frac{I}{I_o} = \frac{I_o - I}{I_o} = 1 - 10^{-\epsilon C \ell} \qquad (1.7)$$

However, if the fluorescence has a quantum yield ϕ_f, the fraction of the absorbed light which appears as fluorescence (I_f/I_a)
is

$$\frac{I_f}{I_o} = \phi_f \frac{I_a}{I_o} = \phi_f (1 - 10^{-\epsilon C \ell}) \qquad (1.8)$$

or in arbitrary units, the intensity of light emitted is

$$I_f = \phi_f I_o (1 - 10^{-\epsilon C \ell}) \qquad (1.9)$$

the term $1 - 10^{-\epsilon C \ell}$ can be expanded in a power series

$$1 - 10^{-\epsilon C \ell} = 2.3\epsilon C \ell + \frac{(2.3\epsilon C \ell)^2}{2!} - \frac{(2.3\epsilon C \ell)^3}{3!} + \frac{(2.3\epsilon C \ell)^4}{4!} - $$

$$(1.10)$$

which reduces to

$$1 - 10^{-\epsilon C \ell} = 2.3\epsilon C \ell \qquad (1.11)$$

if ϵC is very small. Thus, in the limit of low absorber concentration

$$I_f = 2.3\phi_f I_o \varepsilon C\ell \qquad\qquad (1.12)$$

It can be seen then, that on a molecular basis, I_f depends upon
the concentration and molar absorptivity of the absorbing
(ground state) species and the quantum yield of fluorescence,
a property of the fluorescing (excited) species. For analytical
purposes, Eqn. (1.11) has too many factors to determine to be
useful. However, under conditions of constant excitation,
$\phi_f I_o$, ε and I are the same for unknown and standard samples,
and relative fluorimetry can be employed by dividing Eqn. (1.11)
for the unknown sample by Eqn. (1.11) for the standard sample,

$$\frac{I_{fu}}{I_{fs}} = \frac{C_u}{C_s} \qquad\qquad (1.13)$$

where I_{fu} and I_{fs} are the experimentally determined fluorescence
intensities of unknown and standard samples, respectively, and
C_u and C_s are the concentrations of fluorescing material in
unknown and standard samples, respectively.

Phosphorescence

The radiative transition from the lowest triplet state to
the ground singlet state is longer lived than fluorescence
because of the spin forbiddenness (low probability) of the
former and is called phosphorescence. Phosphorescence is, like
fluorescence, most likely to occur in molecules having restricted vibrational freedom and is thus most often observed in aromatic molecules and their derivatives. However, all aromatic
molecules are not phosphorescent. Because triplet states are
so long lived (10^{-5} - several seconds) chemicals and physical
processes in solution as well as internal conversion compete
effectively with phosphorescence for deactivation of the lowest
excited triplet state. Except for the shortest lived phosphorescences, collisional deactivation by solvent molecules,
quenching by paramagnetic species, photochemical reactions, and

energy transfer processes preclude the observation of phosphorescence in fluid media. Rather, phosphorescence is normally
studied in the glassy state at liquid nitrogen temperature, in
solution in very viscous liquids where collisional processes
cannot completely deactivate the triplet state and in low
pressure gases.

Phosphorescence usually originates from the lowest vibrational level of the lowest triplet state and terminates in any
of several vibrational levels of the ground state. Thus the
structure of the phosphorescence band represents the vibrational
structure of the ground state and the separations between vibronic features in the phosphorescence spectrum can usually
be matched with peaks in the infrared or Raman spectra of the
phosphorescing molecule, corresponding to normal vibrations of
the ground state molecule.

The lowest frequency vibrational feature of the phosphorescence spectrum is generally taken to be the 0-0 band of phosphorescence. Unlike the 0-0 band of fluorescence, which ideally
coincides with the 0-0 band of the lowest frequency absorption
maximum (but actually occurs at slightly lower frequency than
the absorption 0-0 band at low temperatures), the 0-0 band of
phosphorescence occurs at substantially lower frequencies than
the 0-0 absorption band. This is a consequence of the lowest
triplet state always lying below the lowest excited singlet
state as a result of the smaller repulsive energy in the triplet
state. It is also the lower electrostatic energy in the lowest
triplet state that prevents intramolecular charge transfer
stabilization of the lowest triplet state, to the same degree
it occurs in the lowest excited singlet state by electron
donor and acceptor substituent groups. As a result, exocyclic
groups do not as efficiently extend the conjugation of the aromatic system in the lowest triplet state as they do in the
lowest excited singlet state and the same substituents tend to
shift the phosphorescence spectra to a smaller extent than
they do the absorption or fluorescence spectra. However, the

extension of aromaticity by annelation has a comparable effect
on the energies of all excited states and therefore a comparable
shifting effect on phosphorescence, fluorescence and absorption.
For purposes of comparison, the 0-0 bands of phosphorescence of
benzene, naphthalene and anthracene lie at 2.95×10^4 cm^{-1},
2.13×10^4 x cm^{-1} and 1.49×10^4 cm^{-1}.

The quantum yield (quantum efficiency) of phosphorescence
(ϕ_p) is defined by

$$\phi_p = \phi_{st} \frac{k_p}{k_p + \Sigma k_j} \qquad (1.14)$$

where k_p is the molecular probability (or macroscopic rate
constant) of phosphorescence, Σk_j is the sum of the rate constants of all unimolecular radiationless deactivation processes
competing with phosphorescence for deactivation of the lowest
triplet state (actually only radiationless intersystem crossing
to the ground state - analogous to spin forbidden internal conversion - competes with phosphorescence in truly isolated molecules), and ϕ_{st} is the efficiency of singlet-triplet intersystem
crossing from the lowest excited singlet state. It should be
noted that ϕ_p represents the fraction of the total number of
molecules originally excited to the lowest excited singlet
state that ultimately phosphoresces and is thus a three state
quantum yield (in contrast to the two state ϕ_f). ϕ_p is defined
with the assumption that the population of the lowest triplet
state by direct singlet-triplet absorption is negligible compared with the efficiency of population of the lowest triplet
from the lowest excited singlet state. If this were not so,
the expression for ϕ_p would become considerably more complicated. High values of ϕ_p are favored by the presence of heteroatoms or substituent atoms of high atomic number (heavy
atoms) because these substituents produce spin-orbital coupling
which favors spin forbidden transitions. Thus, quinoline and
1-bromonaphthalene both phosphoresce more intensely and
fluoresce more weakly than naphthalene. However, because ϕ_{st},
k_p and Σk_j are all increased by spin-orbital coupling, if Σk_j

is increased much more than ϕ_{st} and k_p, it is possible to populate the lowest triplet state efficiently, at the expense of the population of the lowest excited singlet state and still diminish the phosphorescence yield by introduction of a heteroatom or heavy atom. Thus, in many chelating ligands, coordination by very heavy ions such as Hg(II) and TI(III) results in a decrease in both ϕ_f and ϕ_p. Also, while benzene exhibits both fluorescence and phosphorescence, pyridine, having a nitrogen atom in the ring, exhibits neither.

Delayed Fluorescence

Occasionally, in rigid and viscous media, a second long lived emission band is observed in addition to the phosphorescence. The second band is of higher frequency than the phosphorescence, and if the molecule also shows fluorescence at the temperature of measurement, the frequency of the extraneous band coincides with the frequency of fluorescence, but its decay time is similar to that of the phosphorescence. This phenomenon is known as delayed fluorescence and is a result of the thermal excitation of a molecule in the lowest excited triplet state back to the lowest excited singlet state (i.e. reverse intersystem crossing) followed by fluorescence. In rigid media, the thermal excitation is accomplished by warming the sample, or by statistical fluctuations in the thermal energy of the triplet state. However, in viscous media where there is some mobility, the collision of two triplet molecules results in the thermal excitation of one, and the internal conversion of the other to the ground state. The former process is called E-type delayed fluorescence (because it was first observed in eosin) and is exhibited only by certain molecules which have a small energy gap between the lowest excited singlet and triplet states, making thermal repopulation of the lowest excited singlet state probable. The latter process is called triplet-triplet annihilation or P-type delayed fluorescence (because it was first observed in pyrene), and may occur even in molecules with substantial differences in energy between the lowest excited singlet and triplet states. The total energy available

for the process is the energy of the triplet state, which is
generally more than enough to promote one triplet state molecule to the lowest excited singlet state. Because two triplet
molecules are required to produce one P-type delayed fluorescence event, fairly high concentrations of the absorbing species
are usually required. P-type delayed fluorescence is a bipho-
tonic process, because the absorption of two photons (or waves)
of exciting light must be absorbed to produce one event. Thus,
the intensity of P-type delayed fluorescence is proportional
to the square of the intensity of exciting light. E-type delayed fluorescence, however, is monophotonic, and the intensity
of this type of delayed fluorescence varies linearly with the
first power of the intensity of exciting light.

It is important to be aware of the possibility of occurrence of delayed fluorescence, because the appearance of a
second long lived band may lead to the erroneous conclusion
that either an impurity or a double phosphorescence (from one
species) is being seen. The actual utility of delayed fluorescence in analytical and biological science has yet to be
developed.

BIBLIOGRAPHY

Jaffe, H. H. and Orchin, M., _Theory and Applications of Ultra-violet Spectroscopy_, Wiley, New York, 1962.

Kasha, M., _Discussions Faraday Soc._, 9, 14 (1950).

Klevens, H. B. and Platt, J. R., _J. Chem. Phys._, 17, 470 (1949).

Murrell, J. N., _Theory of the Electronic Spectra of Organic Molecules_, Wiley, New York, 1963.

Parker, C. A., _Photoluminescence of Solutions_, American Elsevier, New York, 1967.

Sidman, J., _Chem. Revs._, 58, 689 (1958).

Turro, N., _Molecular Photochemistry_, Benjamin, New York, 1965.

Winefordner, J. D., Schulman, S. G. and O'Haver, T. C., _Luminescence Spectrometry in Analytical Chemistry_, Wiley, New York, 1972.

CHAPTER 2
PHOTOPHYSICAL PROCESSES
IN MOLECULES IN SOLUTION

In the preceding chapter, the arguments developed concerning the spectroscopic behavior of molecules were based upon the intrinsic electronic properties of the molecules themselves. This treatment is complete only for truly isolated and noninteracting molecules, and is realized only in the gas phase at low pressures, and to a fair approximation in dilute solutions in nonpolar and nonhydrogen bonding solvents (e.g. hydrogen solvents). However, most electronic spectroscopy is practiced in fluid or rigid solutions involving solvents, many of which are highly polar or capable of hydrogen bonding with the absorbing or emitting molecules. The interactions of solute molecules with polar or hydrogen bonding solvents are capable of profoundly altering the electronic properties of the states from which absorption and emission occur, and any attempt to account for electronic spectral phenomena in solution must take this into account. Fortunately, the effect of solvents on electronic spectra can be treated in terms of the perturbation of the spectra of the isolated molecules by the interaction with the solvent. This will allow us to build upon the theory already developed for the spectra of isolated molecules. Moreover, in activating (polar and hydrogen bonding) solvents, interactions between solute molecules (e.g. chemical reactions) which were kinetically or thermodynamically impossible in isolated molecules without solvent assistance often occur in ground and electronically excited states. These interactions usually have profound effects upon the electronic spectra observed and must be taken into account in order to explain spectral behavior in condensed media. In this chapter we shall consider the effects of solvent-solute and solute-solute interactions upon the properties of $\pi \to \pi^*$, $n \to \pi^*$ and intramolecular charge transfer transitions and the spectral bands they represent.

Solvent Effects on Electronic Spectra

Solvent interactions with solute molecules are predominantly electrostatic and may be of the induced dipole-induced
dipole, dipole-induced dipole, dipole-dipole or hydrogen bonding types (1). Additionally, hydrogen bonding usually accom-
panies dipole-dipole interaction as a mode of solvation. A
solvent which has positively polarized hydrogen atoms which can
engage in hydrogen bonding is said to be a hydrogen bond donor
solvent. A solvent which has atoms having lone or nonbonding
electron pairs is said to be a hydrogen bond acceptor solvent.
Qualitatively, a hydrogen bond donor behaves as a very weak
Bronsted acid, partially donating a proton to a basic site on
the solute molecule. A hydrogen bond acceptor behaves as a very
weak Bronsted base, partially accepting a proton from the solute
molecule. Because of the involvement of nonbonding and lone
pairs in $n \rightarrow \pi^*$ and intramolecular charge transfer transitions,
hydrogen bonding solvents have the greatest effect on the
positions of these types of spectra. Because of the large
dipole moment changes accompanying electronic reorganization
in $\pi \rightarrow \pi^*$ and intramolecular charge transfer transitions, these
types of spectra are most affected by solvent polarity. Let us
now consider these phenomena in some detail.

Consider the absorption of light by a polar molecule capable of hydrogen bonding, in a solvent of high polarity (dielectric strength) and having both hydrogen bond donor and acceptor
properties (e.g. water). In the ground state the molecule will
have a solvent cage in which the positive ends of the solvent
dipoles will be oriented about the negative ends of the solute
dipole and the negative ends of the solvent dipoles will be
oriented about the positive ends of the solute dipole. Positively polarized hydrogen atoms of the solvent may be oriented
toward lone pairs on the solute and acidic hydrogen atoms of
the solute may be oriented toward lone pairs on the solvent
(Fig. 2.1). The solvent cage is in thermal equilibrium with
the ground state electronic distribution of the solute. The
light absorption process alters the electronic distribution of

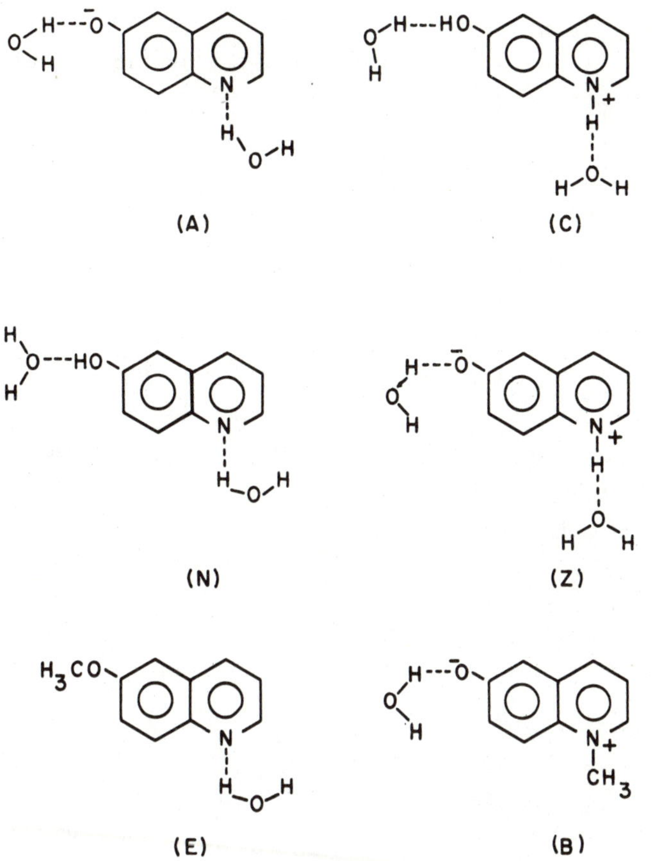

Fig. 2.1. Illustrating possible models of hydrogen bonding
and solvent dipole orientation in various species
derived from 6-hydroxyquinoline in water. C, A, N,
Z, E and B are the cation, anion, neutral molecule,
zwitterion, methyl ether and methyl betaine species,
respectively.

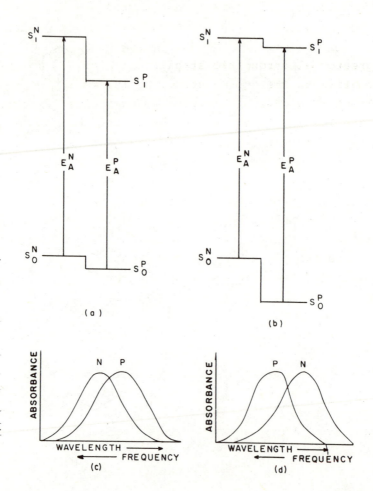

Fig. 2.2. The effect of going from a nonpolar solvent (denoted by N) to a polar solvent (denoted by P), upon the energy (E_A) of an absorptive transition (a) when the excited singlet state to which absorption occurs (S_1) is more polar than the ground state (S_0), (b) when the excited singlet state to which absorption occurs is less polar than the ground state, (c) and (d) represent the way the spectral bands corresponding to the transitions in (a) and (b), respectively, might appear.

the solute so that the electronic dipole moment of the excited
molecule is different from that of the ground state molecule.
However, the absorption process is so rapid that it terminates
with the excited molecule still in the ground state equilibrium
solvent cage (i.e. in a Franck-Condon excited state). If the
solute molecule becomes more polar in the excited state, there
will be greater electrostatic stabilization of the excited
state, relative to the ground state, by interaction with the
polar solvent (Fig. 2.2). The greater the polarity of the
solvent, the lower will be the energy of the Franck-Condon
excited state. This type of behavior is characteristic of most
$\pi \rightarrow \pi*$ and intramolecular charge-transfer transitions and is
observed as a shift to longer wavelengths of the absorption
band with increasing solvent polarity. In the event that the
electronic dipole moment is lower in the Franck-Condon excited
state than in the ground state, increasing solvent polarity
will stabilize the ground state to a greater degree than the
excited state and the absorption spectrum will shift to shorter
wavelengths with increasing solvent polarity.

Absorptive transitions of the $n \rightarrow \pi*$ type are usually more
affected by hydrogen bond donor properties of the solvent than
by solvent polarity per se. If a nonbonding pair on a solute
molecule is bound by a hydrogen atom of the solvent, the hydrogen bonding interaction stabilizes the ground state as well as
the $n,\pi*$ state of the solute. However, because the ground state
molecule has two electrons in the nonbonding orbital and the
excited state has only one, the stabilization of the ground
state is greatest. As a result, the energies of $n \rightarrow \pi*$ absorptions increase (the spectra shift to higher frequencies or
shorter wavelengths) with increasing solvent hydrogen bond donor
capacity. In a similar manner, but to a lesser degree, the
positive end of the dipole of a polar solvent is capable of
producing the same effect upon $n \rightarrow \pi*$ absorption spectra. Hydrogen bonding solvents also produce a dramatic effect upon
intramolecular charge-transfer absorption spectra. Hydrogen
bond donor solvents interacting with lone pairs on functional

groups which are charge-transfer acceptors in the excited state
(e.g. the carbonyl group), enhance charge-transfer by introducing a partial positive charge into the charge-transfer acceptor
group. This interaction stabilizes the charge-transfer excited
state relative to the ground state so that the absorption spectra shift to longer wavelengths with increasing hydrogen bond
donor capacity of the solvent. Following similar lines of
reasoning, increasing hydrogen bond donor capacity of the solvent produces shifts to shorter wavelengths when interacting
with lone pairs on functional groups which are charge-transfer
donors in the excited state (e.g. -OH, -NH$_2$). Hydrogen bond
acceptor solvents produce shifts to longer wavelengths when
solvating hydrogen atoms on functional groups which are charge-
transfer donors in the excited state (e.g. -OH, -NH$_2$). This is
effected by the partial withdrawal of the positively charged
proton from the functional group. Finally, solvation of hydrogen atoms on functional groups which are charge-transfer accep-
tors in the excited state (e.g. $-\overset{O}{\overset{\|}{C}}-OH$), inhibits charge-transfer
by leaving a residual negative charge on the functional group.
Thus, the latter interaction results in shifting of the absorption spectrum to shorter wavelengths.

In fluorescent molecules, subsequent to excitation to the
Franck-Condon excited state, the ground state solvent cage re-
orients itself to conform to the new electronic distribution of
the excited molecule. This solvent relaxation process involves
reorientation of solute dipoles about new centers of positive
and negative charge in the excited molecule, and possibly the
strengthening, weakening, breaking and making of hydrogen bonds.
Because nuclear motions are involved, solvent relaxation is
approximately contemporaneous with vibrational relaxation,
taking about 10^{-14} - 10^{-12} sec, and is rapid by comparison with
the lifetime of the lowest excited singlet state ($\sim 10^{-8}$ sec).
Consequently, fluorescence originates from the excited solute
molecule in a thermally equilibrated solvent cage configuration
which is lower in energy than the Franck-Condon excited state,
and generally even somewhat lower than the vibrationally relaxed

unsolvated or weakly solvated excited molecule. When fluorescence occurs, it terminates in the ground electronic state of
the solute molecule, but because of the rapidity of the electronic transition, the molecule is still in the excited state
equilibrium solvent cage (higher in energy than the thermally
relaxed ground state). Rapid solvent relaxation then occurs
($10^{-14} - 10^{-12}$ sec), and the solute molecule ultimately returns
to the ground state equilibrium solvent cage. Because the solvent relaxed excited state is lower in energy than the Franck-
Condon excited state, and the Franck-Condon ground state is
higher in energy than the solvent relaxed ground state (Fig.
2.3), fluorescence often occurs at considerably longer wavelengths than would be anticipated purely on the basis of
vibrational relaxation. It is for this reason that the 0-0
bands of fluorescence and absorption often do not coincide.

Solvent polarity and hydrogen bonding effects upon fluorescence spectra are qualitatively similar to those upon absorptive
spectra. For example, molecules which are more polar in the
excited state usually show longer wavelengths of fluorescence
in the more polar solvents. However, molecular sites of
hydrogen bonding in the ground state solvent cage may be
altered in the excited state solvent cage. Since spectral
shifts to longer or shorter wavelengths resulting from dipolar
or hydrogen bonding interactions may be constructively or de-
structively additive, it is not unusual to find that in a given
series of solvents, the fluorescence shifts of a given solute
may not parallel, even qualitatively, the absorption shifts in
the same series of solvents (2). For example, n $\rightarrow$ π^* absorptions almost always shift to substantially shorter wavelengths
with increasing solvent hydrogen bond donor capacity. However,
there appears to be very little regularity in the shifting with
increasing solvent hydrogen bond donor capability for the
limited number of n $\rightarrow$ π^* fluorescences, which have been studied
in different solvents (3).

Absorption spectra do not reveal relaxation processes

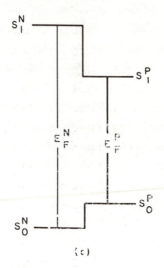

(c)

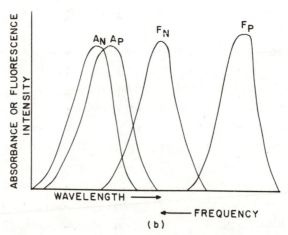

(b)

Fig. 2.3. (a) The effect of solvent upon the energy of the
fluorescent transition (E_F) of a molecule which
is more polar in the excited state (S_1) than in
the ground state (S_0). Superscript N denotes
nonpolar solvent; superscript P denotes polar
solvent.

(b) Representation of the fluorescence bands F_N and
F_P which might be expected to correspond to the
transitions depicted in (a). The corresponding
$S_1 \leftarrow S_0$ absorption bands (A_N and A_P) are also
shown, illustrating the destruction of the mirror
image relationship between absorption and fluorescence in polar solvents.

occurring subsequent to absorption (e.g. solvent relaxation in
the excited state). Similarly, fluorescence spectra do not reveal relaxation processes occurring subsequent to fluorescence
(e.g. solvent relaxation in the ground state). In molecules
which become more polar in the excited state, solvent relaxation
stabilization is greater in the excited state than in the ground
state. Thus, the fluorescence spectra of these molecules tend
to show greater wavelength dependences on solvent polarity than
do the absorption spectra. On the other hand, molecules which
become less polar in the excited state usually show greater
dependence of the absorption spectra on solvent polarity.

The dependences of absorption and fluorescence spectra
upon hydrogen bonding properties of the solvent are somewhat
more involved than the dependence upon solvent polarity alone.

In hydrogen bond donor solvents, lone electron pairs on
aromatic ring substituents (e.g. in phenols and arylamines) are
partially inhibited from excited state charge-transfer interaction with the aromatic ring, resulting in a high frequency
shift of the electronic spectra relative to the spectra in
hydrocarbon solvent. However, charge-transfer from the aromatic
ring to electron acceptor groups (e.g. in aromatic carbonyl
compounds) is enhanced by interaction with hydrogen bond donor
solvents, resulting in a shift to lower frequency of the absorption and fluorescence spectra relative to the spectra in
hydrocarbons. Hydrogen bond acceptor solvents favor conjugation
of lone pairs with the aromatic ring in amines and phenols, and
inhibit conjugation in carboxyl groups resulting in shifts to
lower and higher frequency, respectively, of the electronic
spectra. In many cases, functional groups which are not coplanar with the aromatic ring in the ground state become coplanar (rehybridize) in the excited state, subsequent to excitation, because intramolecular charge transfer is facilitated
by coplanarity; conjugation being most effective when the lone
pair orbitals of the substituent have the same spatial direction
as the p_{π} orbitals of the aromatic ring. Solvent relaxation in

the excited state includes the reorientation and changes in
strength of hydrogen bonds subsequent to excitation, as well as
reorientation of solvent dipoles about the excited molecule.
In cases where hydrogen bonding favors the intramolecular
charge-transfer interaction, the fluorescence spectrum usually
shows greater shifts to longer wavelengths than the absorption
spectrum upon going from nonhydrogen bonding to hydrogen
bonding media. In cases where hydrogen bonding inhibits intramolecular charge-transfer or rehybridization in the excited
state, the absorption and fluorescence spectra usually demonstrate comparable dependences on the solvent hydrogen bonding
capabilities. It must be borne in mind, however, that hydrogen
bonding donor and acceptor capabilities, as well as solvent
polarity, will all affect the actual spectral shifts and the
distinction of the various effects is usually established only
after many painstaking measurements of spectra in solvents of
varying hydrogen bonding and dielectric properties (2,4). In
this regard, the freezing of the solutions may also be extremely
useful. Freezing "locks" the fluorescing molecule into the
ground state equilibrium solvent cage and molecular conformation
and thereby prevents solvent cage relaxation and functional
group rehybridization subsequent to excitation. Thus, the
comparison of absorption and fluorescence spectra taken in
frozen solutions with those taken in fluid media permits distinction between solvent and conformational effects arising from
the ground state circumstances of the molecule and those arising
from the extra-electronic circumstances of the thermally equilibrated excited molecule (5).

There has been little systematic investigation of the frequency dependences of phosphorescence spectra on solvent polarity and hydrogen bonding capability. However, the limited
amount of work which has been done suggests that phosphorescence
spectra show much smaller wavelength variations with solvent
properties than do absorption or fluorescence spectra. To
understand this, it is necessary to consider the electronic
properties of the triplet state. Triplet state acid-base

dissociation studies (4) have shown that the electronic dipole
moments of molecules in the triplet state are only slightly
different from those of the ground singlet state. The similar-
ities of triplet state dissociation constants to ground state
dissociation constants suggest that hydrogen bonding properties
of triplet molecules are not very different from those of the
ground state molecules. Moreover, intramolecular charge-transfer phenomena frequently observed in the lowest excited singlet
state have not been detected for triplet states of molecules
having functional groups likely to engage in charge-transfer.
By contrast, the lowest excited singlet state usually has very
different dipolar and hydrogen bonding properties from that of
the ground state of the same molecule. These observations can
be rationalized on the basis of the greater degree of electron
correlation in the triplet state relative to the excited singlet
state. Because of the identity of spin of the unpaired electrons in the triplet molecule, they must remain farther apart
than those of the excited singlet molecule. Hence, there is
less repulsion energy in the triplet species, and the difference
in electrostatic energy of the triplet is lower than that of
the excited singlet. This accounts for the lower energy of the
triplet state, which results in the wavelength of phosphorescence being lower than that of fluorescence.

The lower repulsive energy of the triplet state also ex-
plains the lower electrostatic reactivity of the triplet molecule, relative to that of the excited singlet state of the same
molecule. It must be borne in mind that phosphorescence is
usually observed only in rigid solutions at low temperatures,
so that valid comparisons with absorption and fluorescence
spectra must be made under the same conditions where solvent
relaxation is impossible. However, measurements of triplet
state dissociation constants in fluid solutions by flash photolysis techniques (4) have yielded essentially identical results
with the same quantities obtained from phosphorescence spectra
in rigid media, and suggest that solvent relaxation phenomena
are not appreciable in magnitude for molecules in the lowest
triplet state.

Up to the present, we have been concerned exclusively with
the effects of solvent polarity and hydrogen bonding properties
on the properties (frequencies) of the electronic spectra bands.
However, the intensities of absorption, fluorescence and phosphorescence also demonstrate dependences upon the nature of the
solvent, although these are not as well understood as the relationship of spectral frequencies to solvent.

The intensities of absorption spectra in different solvents
are dependent upon the ways in which the interactions of the
ground and Franck-Condon excited singlet states of the solute
with the solvent affect the molar absorptivities of the various
absorptive transitions. In functional derivatives of aromatic
ring structures, the two lowest frequency $\pi \to \pi^*$ bands are
forbidden because of angular momentum considerations (see Chapter 1), and are usually low in intensity. Strong interaction of
functional groups on the aromatic rings with the solvent, especially when charge-transfer between the aromatic rings and
functional groups occurs in the excited state, removes the
forbiddenness of these bands by distorting the π-electron systems so that the selection rules, which are based on the symmetrical π-electron distributions of the unsubstituted parent
hydrocarbon, no longer are valid. This results in increased
absorptivity relative to that in the unsubstituted aromatic
ring. Usually dipole-dipole or hydrogen bonding interactions
between the solvent and the solute, which favor intramolecular
charge-transfer interaction in the excited state, result in
further removal of forbiddenness and further hyperchromism of
the ^1L_a and ^1L_b absorption bands. Those solvent interactions
with functional groups which inhibit excited state intramolecular charge-transfer, usually cause decreases in absorptivity of the ^1L_a and ^1L_b bands. Thus, those solvents which
cause shifts to longer wavelengths of the ^1L_a and ^1L_b bands
also cause hyperchromism, while those solvents which cause
shifts to shorter wavelengths cause hypochromism of the ^1L_a
^1L_b bands. The ^1L_b band is the most forbidden by the angular
momentum selection rule for aromatic hydrocarbons, and thus

shows the most dramatic intensity dependence upon the solvent. The ^1B_b band of benzenoid and naphthalenoid compounds is allowed by the angular momentum selection rule and is therefore affected only slightly, by changes in solvent. The absorption bands of aromatic hydrocarbons and their derivatives substituted with nonconjugated functional groups are similarly only slightly affected by solvent properties.

The intensities of fluorescence and phosphorescence spectra are extremely sensitive to solvent polarity and hydrogen bonding properties. These sensitivities to solvents are in part the result of changes in transition intensity produced by distortion of the π-electron distributions of the luminescing molecules and are thus analogous to the effects of solvent perturbations of the electronic absorption spectra. However, the greatest solvent influences on the intensities of the luminescence transitions (as reflected through solvent dependences of the quantum yields of fluorescence and phosphorescence in different solvents) are derived from solvent-induced changes in energy and mechanisms of deactivation of excited singlet and triplet states.

Hydrogen bonding in the ground state and electrostatic stabilization of the lowest excited singlet $\pi,\pi*$ state of molecules whose lowest singlet states are $n,\pi*$ in the isolated molecule or in hydrocarbon solvents, raises the energy of the lowest singlet $n,\pi*$ state and lowers the energy of the lowest $\pi,\pi*$ singlet state, respectively. If the stabilization of the $\pi,\pi*$ state and the destabilization of the $n,\pi*$ state are both large, the $\pi,\pi*$ state may become the lowest excited singlet state, favoring the occurrence of fluorescence at the expense of the population of the triplet state. Thus, in molecules with a $^1n,\pi*$ state as the lowest excited singlet state, in the isolated molecule (e.g. quinoline (7) and pyrene-3-aldehyde (8)), fluorescence is favored by high polarity and high hydrogen bond donor capacity of the solvent. Phosphorescence on the other hand is favored by nonpolar, nonhydrogen bonding media in which the lowest $^1n,\pi*$ state remains as the lowest excited singlet state.

In many molecules whose lowest excited singlet states are
of the π,π^* or intramolecular charge-transfer types, increases
in solvent hydrogen bonding or polarity are frequently found
to produce decreases of fluorescence quantum yields (2,4,9).
Moreover, in some solvents in which fluorescence does not occur
under fluid conditions, fluorescence may be observed upon
freezing of the solution. These observations suggest that
hydrogen bonding or dispersion interactions resulting from the
equilibrium excited state solvent cage are responsible for the
low yields of fluorescence in activating solvents, because
freezing of the solvent prevents solvent relaxation in the excited state and confines the solute to the ground state solvent
cage while decreases in solvent activating strength result in
weaker interactions in both ground and excited states. Although
it is certain that hydrogen bonding or dipole-dipole interaction
in the excited state apparently can increase the rate of at
least one nonradiative pathway competitive with fluorescence,
for the deactivation of the lowest excited singlet state, the
detailed nature of the variation of fluorescence quantum yield
with solvent polarity and hydrogen bonding strength remains
largely a mystery. In some cases fluorescence efficiency varies
only with hydrogen bonding properties of the solvent. In some
cases variation is with solvent polarity and in other cases
dependence on both solvent properties is observed. Sometimes
the phosphorescence efficiency increases as the fluorescence
efficiency decreases with solvent activating strength and sometimes not. Moreover, in some cases the influence of solvent
activating strength on fluorescence efficiency is not continuous
and fluorescence may increase and then decrease with increasing
hydrogen bonding strength (4). The study of the mechanisms
whereby fluorescence and phosphorescence quantum yields vary
with solvent polarity and hydrogen bonding properties should
provide a fertile area for future research.

There is one type of solvent effect upon electronic spectra, known as the external heavy atom effect, which is not related to polarity or hydrogen bonding and which affects the

intensities of fluorescence, phosphorescence and certain absorptive transitions but which does not have an appreciable effect
on the frequencies of the transitions. In solvents whose molecules contain atoms of high atomic number, such as iodine (e.g.
ethyl iodide and diiodo methane), the high nuclear charge of the
heavy atoms in solvent molecules of the primary solvent cage of
the solute molecule causes the spin and orbital angular momenta
of the solute to interact strongly with each other. The interaction of the spin and orbital electronic motions which is very
weak in the absence of heavy atoms, results in the loss of the
total molecular spin angular momentum as a well defined molecular property. As a result, the selection rule forbidding
changes of spin angular momentum in electronic transitions is
partially removed with the consequence that the probabilities
of singlet-triplet absorptive transitions, singlet-triplet
intersystem crossing from the lowest excited singlet state,
phosphorescence and triplet-singlet intersystem crossing from
the lowest triplet to the ground state are substantially increased. Consequently, in some aromatic molecules dissolved in
heavy atom solvents, weak absorption bands ($\varepsilon < 0.1$) at lower
frequencies than the lowest frequency singlet-singlet absorption
bands become apparent. The lowest singlet-triplet absorption
band lies at lower frequency than the lowest singlet-singlet
absorption band because the lowest triplet state is lower in
energy than the singlet state of the same electronic orbital
configuration. Molecules which fluoresce in nonheavy atom solvents generally fluoresce less strongly or may even be completely
quenched in heavy atom solvents because the higher probability
(greater rate) of singlet-triplet intersystem crossing in the
heavy atom solvent results in greater population, per unit time,
of the lowest triplet state at the expense of the population of
the lowest excited singlet state. Thus fewer molecules are
capable of fluorescing at any instant of time. The high population of the triplet state produced by the external heavy atom
favors phosphorescence and often phosphorescence yields are
enhanced in heavy atom solvents, a fact of considerable importance in the choice of a solvent in which to perform analytical

phosphorimetry. However, the external heavy atom effect also
increases the rate of radiationless as well as radiative deactivation of the lowest triplet state and occasionally, the
radiationless process is favored over phosphorescence. Consequently, it is not unusual to find that heavy atom solvents
decrease the yields of phosphorescence as well as fluorescence.
Because of the greater probability per unit time of singlet-
triplet processes in heavy atom solvents the lifetimes of triplet
states, as determined from the rates of decay of phosphorescences
in these solvents are generally shorter than in nonheavy atom
solvents.

Acidity Effects on Electronic Spectra

Addition of acids or bases to the solvents in which absorbing, fluorescing or phosphorescing molecules (having functional
groups with dissociable protons or lone or nonbonded electron
pairs) are studied, can affect the electronic spectra in two
ways. If the acidity of the medium after addition of acid or
base is insufficient to protonate lone or nonbonded electron
pairs or to abstract a proton from a dissociable group, the
acid or base may form hydrogen bonds with the basic or acidic
groups of the molecule of spectroscopic interest. The effects
of these types of hydrogen bonding upon the electronic spectra
are similar to those described for hydrogen bond donor and
acceptor solvents in the previous section. Thus the additions
of small amounts of trifluoroacetic acid to solutions of aromatic carboxylic acids in hydrocarbon solvents or of small
amounts of 1,4-dioxane to solutions of phenols in hydrocarbon
solvents produce shifts to longer wavelengths of the absorption
and fluorescence spectra, relative to the electronic spectra
in pure hydrocarbon solvents. In some cases aromatic molecules
having nonbonded electron pairs fail to fluoresce in non-
activating solvents because the lowest excited singlet state is
of the n,π^* type and favors radiationless intersystem crossing
as a mode of deactivation of the lowest excited singlet state.
The addition of small amounts of acid results in hydrogen

bonding with the nonbonded pairs often raising the energy of the
n,π* to such a degree that the lowest π,π* state becomes the
lowest excited singlet state, making fluorescence likely. In
this regard, several nitrogen heterocyclics such as quinoline
and acridine and some aromatic carbonyl compounds such as 2-
acetonaphthone and pyrene-3-aldehyde do not fluoresce but phosphoresce in hydrocarbon solvents. The addition of small amounts
of acids such as trifluoroacetic acid or trichloroacetic acid
which are fairly soluble in hydrocarbons, results in the appearance of fluorescence and a decrease in the intensity of phosphorescence from these molecules.

If the acidity of the medium after the addition of acid or
base is sufficient to protonate functional groups having lone or
nonbonded electron pairs, or to abstract a proton from acidic
functional groups, the effects on the electronic spectra are
more dramatic but qualitatively similar to the effects produced
by hydrogen bonding. Protonation of a basic molecule, or dissociation of an acidic molecule, produces a chemical species
whose reactivity and electronic structure is different from that
of the original molecule. However, protonation is similar to
interaction with a hydrogen bond donor solvent in that a positive polarizing influence is effected at the protonated functional group while dissociation is similar to interaction with
a hydrogen bond acceptor solvent in that the removal of a
positively charged proton is equivalent to a negative polarizing
influence at the dissociated group. Following the lines of
reasoning developed for hydrogen bonding effects on electronic
spectra, some generalizations can be made. Protonation of nonbonded pairs on functional groups which are charge-transfer
acceptors in the excited state (e.g. carbonyl and carboxyl
groups) enhances the acceptor properties of these groups and
results in stabilization of the excited states relative to the
ground state. Protonation of these types of functional groups
therefore produces shifting of the intramolecular, electronic,
charge-transfer spectra to longer wavelengths. However, the
n,π* states giving rise to n → π* transitions (which are usually

seen only in the absorption spectra) of these molecules are
raised in energy to such an extent by protonation that they
disappear completely from the absorption spectra. Protonation
of lone pairs on functional groups which are charge-transfer
donors in the excited states (e.g. $-NH_2$) inhibits the donor
properties of these groups and results in shifting of the electronic spectra to shorter wavelengths. In functional groups,
such as the amino group which have only one lone pair, protonation effectively converts the lone pairs to a dative σ-bond
resulting in the total unavailability of the lone pair for
charge-transfer interaction. Protonation of molecules containing
this type of functional group results in electronic spectra which
are almost identical to the electronic spectra of the unsubstituted molecules. Protolytic dissociation from functional groups
which are charge-transfer acceptors in the excited state (e.g.
carboxyl groups) leaves residual negative charge on the functional groups, inhibiting charge acceptance from the aromatic
ring and producing a shift to shorter wavelengths of the electronic spectra. Finally, protolytic dissociation from functional
groups which are charge-transfer donors in the excited state
(e.g. phenolic groups) enhances charge donation because of the
repulsive effect of the residual negative charge of the dissociated group upon the lone pair electrons and results in
shifts to longer wavelengths of the electronic spectra upon
dissociation. These spectroscopic arguments are equally valid
for absorption, fluorescence and phosphorescence spectra. However, the chemical dynamics that govern the acid-base reactions
and the attendant spectroscopic changes are somewhat different
for the three types of electronic spectra.

The acidity dependence of molecular electronic absorption.
Absorption spectra originate in the thermally equilibrated
ground electronic state and terminate in a Franck-Condon excited
state but the intensity of absorption (absorbance) is proportional only to the concentration of absorbing species (in the
ground state). Because the ground electronic state is of
infinite lifetime, the kinetics of acid-base reactions are much

too fast (rate constants for acid-base reactions are typically
of the order of 10^{10} sec^{-1} - 10^7 sec^{-1}) to affect the acidity
dependence of electronic spectra. Thus absorption spectroscopy
as it pertains to acid-base reactions always entails spectroscopic measurements upon systems in chemical equilibrium.

Provided that the acidic or basic groups of an aromatic
acid or base are directly coupled electronically, to the aromatic
ring, the absorption spectrum of an acid and that of its conjugate base or of a base and of its conjugate acid, will differ as
described previously. For a constant formal concentration C of
an aromatic acid or base, the concentrations of conjugate acid
and conjugate base, at any pH, in dilute aqueous solutions are
related by the Henderson-Hasselbach equation

$$pH = pK_a + \frac{[B]}{[A]}$$ (2.1)

where [A] and [B] are the equilibrium concentrations (activities)
of the aromatic acid or base in the conjugate base and acid
forms, respectively, the pK_a is the logarithm of the reciprocal
of the dissociation constant of the acid or protonated base. If
the absorbance (α) of a solution containing conjugate acid A and
conjugate base B, both being present in measurable amounts (i.e.
pH = $pK_a \pm 2$) is measured at a fixed wavelength λ, and in a cell
of optical path length ℓ, and A and B are the only species absorbing at wavelength λ, then α is related to [A] and [B] by

$$\alpha = \varepsilon_A[A]\ell + \varepsilon_B[B]\ell$$ (2.2)

where ε_A and ε_B are the molar absorptivities of A and B at wavelength λ.

$$C = [A] + [B]$$ (2.3)

and combination of Eqns. (2.1), (2.2) and (2.3) yields

S. G. Schulman

$$pH = pK_a + \log \frac{(\alpha - C\ell/C\ell)}{(\alpha/C\ell - \varepsilon_B)} \qquad (2.4)$$

which enables calculation of the pK_a of an aromatic acid or base
from the differences between the absorption spectral properties
of the acid or base and its conjugate form. It should be mentioned at this point, that if only a single conjugate acid and
only a single conjugate base derived from a given aromatic molecule are present in solution, there will be at least one wavelength in the absorption spectrum at which no changes in absorbance will occur with changing pH. This wavelength is called an
isosbestic point in the spectrum and is characterized by the
equivalence of molar absorptivities of conjugate acid and base
at that wavelength (i.e. $\varepsilon_A = \varepsilon_B$ at the isosbestic point). Because of the failure of absorbance to change with pH at an
isosbestic point, pK values cannot be evaluated by absorption
spectroscopy with measurements made at an isosbestic point.
However, because of the invariance of absorbance with pH at the
isosbestic point it is sometimes advantageous to perform analytical measurements at isosbestic wavelengths. The presence of an
isosbestic point in the acidimetric or alkalimetric titration of
an aromatic acid or base is characteristic of a two component
equilibrium. Titrations of aromatic acids and bases in one
solvent by stepwise addition of another solvent of different
polarity or hydrogen bonding capability also frequently show
isosbestic points due to the presence of equilibrium mixtures
of two differently solvated forms of the absorbing molecule.
In cases where the solvent properties of solutions of aromatic
acids and bases change during the course of acidimetric or
alkalimetric titrations (e.g. titrations in nonaqueous media)
and in titrations of polyacidic or polybasic molecules which
have several functional groups with similar pK_a values, more
than two distinct absorbing species may be present in solution
during the course of titration, resulting in the absence of
isosbestic points over part or all of the acidity interval of
the titration.

The foregoing arguments are related strictly to the equilibrium species distributions of aromatic acids and bases in
their ground electronic states. Fluorescence and phosphorescence, however, originate from the lowest excited singlet and
triplet states, respectively. Thus the acidity dependences of
the molecular luminescence processes are related to the conjugate acid and base species distributions in the excited singlet
and triplet states. These distributions are in turn dependent
upon the rates and types of processes competing with luminescence for deactivation of the short lived excited states as well
as the efficiency with which the excited states are populated.
We shall now consider, in some detail, the factors which govern
the spectroscopic aspects of acid-base reactions in the electronically excited states.

The pH dependence of molecular fluorescence. The excitation
of a molecule from its ground electronic state to a higher electronic state is accompanied by a change in the electronic dipole
moment of the molecule. Accordingly, the distribution of electronic charge in the excited molecule is different from that in
the ground state. Thus the excited molecule may be considered
to have different chemical properties from those of the ground
state molecule. This is the basis of photochemistry. Ordi-
narily, no matter to which electronic state the molecule is
excited, spin is conserved in the excitation process and rapid
vibrational deactivation carries the excited molecule to the
lowest vibrational level of the lowest excited singlet state
(if the ground state is also a singlet state, as is the case for
most organic molecules). If the vibrational freedom of the molecule is restricted, as in aromatic structures, return to the
ground state may occur, from the lowest excited singlet state,
with the excess energy being released in the form of fluorescence. Vibrational relaxation then returns the molecule to the
lowest vibrational level of the ground state. In almost every
case of fluorescence known, emission originates from the lowest
excited singlet state. Thus it is electronic distribution or
the photochemistry of the lowest excited singlet state which is
of concern here.

Among the chemical properties of the lowest excited singlet
state which are different from those of the ground state are
acidity and basicity of the excited molecule. If the electron
density in the region of an acidic or basic group is lower in
the excited state than in the ground state, the molecule will be
a stronger acid or a weaker base in the excited state. If
charge-transfer to an acidic or basic group takes place upon
excitation, the excited molecule will be a weaker acid or a
stronger base than in the ground state.

The lifetimes of molecules in the lowest excited singlet
state (the average time a molecule spends in the excited state
before fluorescing) are typically of the order of 10^{-9} - 10^{-8}
sec. Typical mean times for protolytic reactions vary from a
few orders of magnitude shorter, to a few orders of magnitude
longer than fluorescence lifetimes. Consequently, excited state
prototropic reactions may be much slower, much faster or competitive with radiative deactivation of the excited molecules. The
variations of fluorescence with solution acidity for any particular conjugate acid-base pair depend upon which of the latter
three circumstances prevails (Fig. 2.4).

Case 1 - Excited state proton transfer is much slower than
fluorescence. In this case, if the acid form of the analyte is
excited in a region where the acid is the stable form of the
molecule in the ground state but the conjugate base is the
stable form in the excited state, the excited acid will fluoresce
before it can convert to the excited conjugate base. Similarly,
if the conjugate base is excited it will fluoresce before it can
accept a proton from the solvent. Thus which species emit fluorescence is governed only by which species absorb exciting
light and the relative intensities of emission from acid and
conjugate base are determined exclusively by the thermodynamics
(pK_a) of the ground state prototropic reaction (i.e. the variations of fluorescence intensity of each species with pH parallel
the variations of absorbance of each species with pH). For
example, the ground state pK_a of 9-anthroic acid as determined

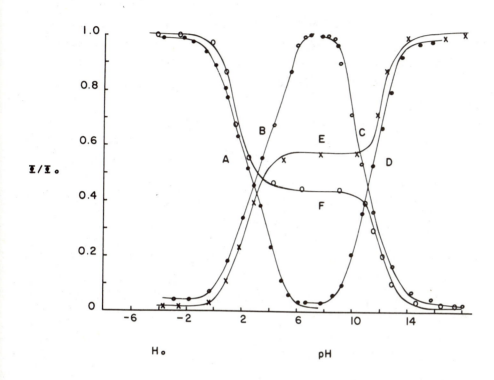

Fig. 2.4. Variations of relative fluorescence efficiency
(fluorescence intensity) with pH of three acids
(A,C and E) and their conjugate bases (B,D and F),
each pair having a ground state pK_a of 11.0 and a
pK_a^* in the lowest excited singlet state of 3.0.
In the conjugate pair A-B, prototropic equilibrium
has been established during the lifetime of the
lowest excited singlet state. In the conjugate
pair C-D, fluorescence of both C and D occurs
much faster than any appreciable amount of excited
state proton exchange. In the conjugate pair E-F,
fluorescence of E and F and proton exchange between
E and F in the lowest excited singlet state, occur
at comparable rates.

by potentiometry or absorptiometry is 3.0. The decrease of the
blue fluorescence of the 9-anthroate anion and the increase of
the green fluorescence of 9-anthroic acid with decreasing pH
also yield a pK_a value of 3.0 (10).

There are, however, some analytical aspects of pH dependent
fluorimetry to be considered, aside from the generally greater
sensitivity of fluorimetry over absorptiometry, in this case.
In the limit of low absorber (analyte) concentration, the intensity of fluorescence I_f, registered on the detector of the
fluorimeter, for each fluorescing species is given by

$$I_f = 2.3\phi_f I_o \alpha \qquad (2.5)$$

where ϕ_f is the quantum yield of fluorescence (the ratio of
excited molecules fluorescing to the total number of molecules
excited), I_o is the intensity of exciting radiation and α is the
absorbance of the test solution at the excitation wavelength.
The greater I_f the more sensitive will be the analysis. Now I_o
is an instrumental constant and α is governed by the molar
absorptivity of the absorbing species at the wavelength of excitation. However, ϕ_f may be different for acid and conjugate
base species derived from the same molecule. In some cases ϕ_f
is zero so that one member of a conjugate pair will be fluorescent while the other is not. Consequently, proper pH adjustment
is highly desirable to obtain maximal analytical sensitivity.

As a result of the difference in electronic distribution in
ground and excited state of the same molecule, the solvent cage
in equilibrium with the ground state molecule is frequently not
in equilibrium with the excited molecule immediately after absorption of radiation. This is especially true in highly polar
and hydrogen bonding solvents. However, the time required for
reorientation of the solvent cage to the equilibrium configuration for the excited molecule is comparable to the time required
for vibrational relaxation (10^{-14} - 10^{-13} sec). Thus solvent
relaxation in the excited state is rapid compared with the

lifetime of the lowest excited singlet state and occurs completely, prior to fluorescence. Solvent relaxation of the excited molecule usually entails a lowering of the energy of the
excited state relative to the ground state. When fluorescence
occurs, the molecule arrives in the ground electronic state in
the equilibrium excited state solvent cage configuration, slightly higher in energy than the equilibrated ground state molecule.
The net effect of all this is to shift the fluorescence band to
considerably lower frequencies than where it would appear relative to the corresponding absorption band in a nonpolar, nonhydrogen bonding or rigid solution. This has been discussed at
length in the previous section. Frequently, it is observed that
the shifts, resulting from solvent relaxation phenomena, are
unequal for acid and conjugate base derived from the analyte;
as for example, in the case of o-phenanthroline and its slightly
protonated conjugate acid (11) the latter of which lies at
anomalously low frequency. This effect can introduce greater
selectivity into fluorimetric analysis by allowing selection of
the fluorescence band most isolated from the emissions of interfering substances for analytical work. Adjustment of the pH of
the test solution to generate the conjugate species giving rise
to the desired fluorescence band is easily accomplished.

Case 2 - Excited state prototropism is much faster than
fluorescence. If the rate of dissociation of the excited acid
and that of protonation of the excited conjugate base is much
greater than the rate of deactivation of the lowest excited
singlet states by fluorescence, prototropic equilibrium in the
lowest excited singlet state will be achieved (12). In this
event it is the thermodynamics of the excited state prototropic
reaction (the dissociation constant of the excited conjugate
acid, pK_a^*) which predominately determines the fluorescence
behavior of the analyte. Since the electronic distribution of
an electronically excited molecule is generally different from
that of its ground state the pK_a^* of the excited acid is
usually very different from the pK_a of the ground state acid.
Differences between pK_a^* and pK_a are commonly six or more

logarithmic units and differences of seventeen or more logarithmic units are not rare. The difference between pK_a^* and pK_a means that the conversion of acid to conjugate base in the excited state occurs in a pH region different from the corresponding ground state reaction. For example, the absorption spectrum of 3-aminoquinoline ($pK_a = -0.4$) at Hammett acidity -0.4 shows the absorption bands originating from both the singly and doubly protonated cations while the fluorescence spectrum consists solely of the emission from the excited singly protonated cation. However, at Hammett acidity -5.6, the absorption spectrum is that of the doubly protonated cation alone while the fluorescence spectrum shows the blue green fluorescence of the singly protonated cation and the blue fluorescence of the doubly protonated cation (pK_a^* -5.6), indicating that the doubly protonated cation is a stronger acid in the lowest excited singlet state than in the ground state (2).

When prototropic equilibrium is attained within the lifetime of the lowest excited state and one or both members of the conjugate pair are fluorescent, the determination of pK_a^* is a relatively simple matter and is analogous to the spectrophotometric determination of a pK_a. A series of buffer solutions each containing the same amount of analyte is prepared. The fluorescence spectrum of each sample is recorded and the positions of the emission maxima of acid and conjugate base noted. A plot of fluorescence intensity of either species as a function of pH or Hammett acidity allows a rapid estimate of pK_a^* as the value of pH or Hammett acidity at the midpoint of the fluorimetric titration curve. However, if both acid and base fluoresce and their fluorescence bands overlap, a correction for the overlap of the fluorescence of the conjugate species must be applied before the fluorimetric titration curve is plotted (see for example Ref. 3) so that the ordinate of the plot corresponds to the fractional concentration of excited acid or base alone. Alternatively, pK_a^* may be obtained from the Henderson-Hasselbach equation in the form

$$pK_a^* = pH - \log \frac{I_A^O - I_A}{I_A - I_B^O} \qquad (2.6)$$

where I_A is the fluorescence intensity at any point in the fluorimetric titration curve at the analytical wavelength of fluorescence (i.e. when pH or Hammett acidity is equal to pK_a^* $\pm$ 2) and I_A^O and I_B^O are, respectively, the limiting fluorescence intensities when pH < pK_a^* - 2 and when pH > pK_a^* + 2 at the analytical wavelength. In the great majority of aromatic acids and bases which demonstrate excited state equilibrium, the fluorimetric titration breaks occur in concentrated acid or base media whose acidities are represented by Hammett acidity scales rather than by pH. This is due to the low availability of protonating (H_3O^+) and dissociating (OH^-) species in the pH region which makes diffusion limited protonation and dissocia- tion unlikely to occur within the lifetime of the excited state. Because solvent effects on fluorescence frequently accompany acidity and basicity effects on fluorescence in concentrated acid or basic media, the emission maxima may shift as the sol- vent composition is changed in order to vary the acidity or basicity of the medium. It is for this reason that fluorimetric determinations of pK_a^* are best performed by scanning the entire fluorescence spectrum of each solution. In this way the solvent- induced shifts of fluorescence bands can be visually distin- guished from acidity-induced shifts, usually with some intelli- gent guesswork, and the most accurate values of pK_a^*, can thus be obtained.

In cases where excited state prototropism is rapid compared with fluorescence, analytical fluorimetry may be subject to serious errors. Direct application of Eqn. (2.5) for analytical purposes is very tedious, requiring the evaluation of several instrumental and spectrochemical parameters and is seldom employed. Most often Eqn. (2.5) is applied to relative fluori- metry in which case the fluorescence intensity of the unknown sample is compared with that of a sample of known concentration of the analyte. Eqn. (2.5) then becomes

$$\frac{I_{fu}}{I_{fs}} = \frac{A_u}{A_s} \qquad\qquad (2.7)$$

where I_{fu} and I_{fs} are the fluorescence intensities of the unknown
solution and the standard sample, respectively, and A_u and A_s
are the absorbances of the unknown solution and the standard
solution, respectively. The absorbance terms, of course, contain
the analyte concentrations. Eqn. (2.7) follows from Eqn. (2.5)
under the assumption that the geometrical arrangement of the
sample and apparatus, the intensity of exciting light and the
quantum yields of fluorescence are the same in both unknown and
standard samples. The geometry of the analytical system and the
intensity of exciting light can, of course, be easily arranged
to be constant. Furthermore, if excitation is not effected at
an isosbestic point in the absorption spectra of acid and
conjugate base it is generally desirable to adjust the pH of
unknown and standard so that only one conjugate species is absorbing the exciting radiation and absorbing as much of it as
possible. For this purpose a knowledge of the ground state pK_a
is indispensible.

The occurrence of excited state prototropism, however, results in a pH dependence of the quantum yield of fluorescence of
both conjugate species in the region where pH $\sim$ pK_a^*. Thus
arbitrary pH adjustment so that pK_a is very different from pH
may leave the test solutions in a pH region in which pH $\sim$ pK_a^*.
This is a rather undesirable situation since if the emission
is divided amongst two species (or partially quenched), analytical sensitivity is sacrificed. If one member of the conjugate
pair is nonfluorescent, fluorimetric analysis may even be erron-
eously deemed unfeasible if the pH is not made sufficiently
acidic or basic to generate the fluorescent conjugate species.
Consequently, it is highly desirable to know the pK_a^* as well
as the pK_a in order to effect proper pH adjustment. Moreover,
if the pH is arbitrarily adjusted to a region in which pH $\sim$
pK_a^*, the fluorescence intensity will be subject to sharp

fluctuations with environmental variations (e.g. CO_2 absorption)
producing small pH changes. Since these variations may not be
identical for both unknown and standard solutions, the employment of Eqn. (2.7) may result in serious errors. In the latter
circumstance, knowedge of pK_a^* and proper pH adjustment is also
critical. However, in some cases, notably in very strong acid
and basic solutions, proper acidity adjustment is not possible
because there is a practical limit to how strongly acidic or
basic a solution can be made. In this case, if both members of
the conjugate pair are fluorescent, the fluorescence spectra
overlap and the limiting quantum yields of fluorescence are
identical, an isoemissive point, analogous to an isosbestic
point in the absorption spectra, will appear in the fluorescence
spectra of acid and conjugate base (13). Fluorescence intensities monitored at an isoemissive point are proportional only
to the total analyte concentration and are not dependent on the
concentrations of either member of the conjugate pair. Thus pH
dependent fluorimetric errors may be avoided.

Excited state prototropism may also work to the advantage
of the analyst, especially in the attainment of analytical
selectivity. Since many molecules are fluorescent in only one
conjugate form, proper pH adjustment, taking account of differences in pK_a^* values may be employed to isolate the fluorescence of a single analyte in a complex sample. Moreover, due
to the fact that charge-transfer effects accompanying excitation
are frequently very large, molecules with different substituents
and having very similar ground state prototropic behavior may
demonstrate very different excited state behavior. For example,
8-quinolinol and many of its substituted derivatives fluoresce
in the excited cation form in concentrated acid solutions but
not in dilute acid (14). 8-Quinolinol itself has a pK_a, determined by potentiometry or spectrophotometry of 5.1. 5-Fluoro-
8-quinolinol has a pK_a of 4.9. However, as determined by fluorimetry, pK_a^* for 8-quinolinol is -7 while that for the 5-fluoro
derivative is -11. Thus it is possible to determine the unsubstituted material by adjusting the Hammett acidity of the

test solution, with sulfuric acid, to -9, observing the fluorescence of the 8-quinolinol alone and then making the solution yet
more acidic to generate the 5-fluoro-8-quinolinolium ion fluorescence and determine the 5-fluoro derivative by difference.
Clearly, such selectivity is not possible with potentiometry or
absorptiometry. Moreover, the pK_a^* is a constant of the fluorescing molecule which can aid in its identification.

Instances where prototropic equilibrium is essentially
complete within the lifetime of the lowest excited singlet state
are often observed. Frequently, however, partial equilibrium
is attained. This circumstance will be treated in the following
section.

Case 3 - Excited state proton transfer and fluorescence
occur at comparable rates. If the rates of proton transfer to
and from excited acid and conjugate base are comparable to the
rates of deactivation of acid and conjugate base, by fluorescence, the variations of the relative quantum yields of fluorescence of acid and conjugate base, with pH, will be governed by
the kinetics and mechanisms of the excited state prototropic
reactions. In this case, the descriptions of the fluorimetric
titrations of acid and conjugate base require specification of
the mechanism of proton transfer in the excited state. In
acidic solutions (pH < 7), with no buffer ions present (i.e.
H_3O^+ or H^+ is the only protonating species and H_2O is the only
proton acceptor present in appreciable quantity), proton exchange occurs according to

$$HA\ (+H_2O) \underset{k_{-a}}{\overset{k_a}{\rightleftharpoons}} H^+\ (or\ H_3O^+)\ +\ A^- \qquad (2.8)$$

In basic solutions (pH > 7), with no buffer ions present (i.e.
H_2O is the only proton donor and OH^- is the only proton acceptor
present), the mechanism of excited state proton exchange is

$$A^- + H_2O \; \underset{k_{-b}}{\overset{k_b}{\rightleftharpoons}} \; HA + OH^- \qquad (2.9)$$

HA can represent either a neutral or charged conjugate acid and A^- either a neutral or charged conjugate base. If buffer ions BH^+ and B are also present in solution, the reaction

$$HA + B \; \underset{k_{BH^+}}{\overset{k_B}{\rightleftharpoons}} \; A^- + BH^+ \qquad (2.10)$$

must be considered in addition to Eqns. (2.8) or (2.9), depending upon the pH of the solution.

Weller (15,16) has employed simple steady state kinetics to show that, to a first approximation, the variations of the relative quantum yields of fluorescence of HA(ϕ/ϕ_o) and A^- (ϕ'/ϕ') with the concentrations of protonating and deprotonating species are given by

$$\phi/\phi_o = \frac{1 + k_{-a}\tau_o'[H^+]}{1 + k_a\tau_o + k_{-a}\tau_o'[H^+]} \qquad (2.11)$$

and

$$\phi'/\phi_o' = \frac{k_a\tau_o}{1 + k_{-a}\tau_o + k_{-a}\tau_o'[H^+]} \qquad (2.12)$$

if mechanism (2.8) is operative;

$$\phi'/\phi_o' = \frac{1 + k_{-b}\tau_o[OH^-]}{1 + k_b\tau_o' + k_{-b}\tau_o[OH^-]} \qquad (2.13)$$

and

$$\phi/\phi_o = \frac{k_b \tau_o'}{1 + k_b \tau_o' + k_{-b} \tau_o [OH^-]} \qquad (2.14)$$

if mechanism (2.9) accounts for excited state proton exchange;

$$\phi/\phi_o = \frac{1 + (k_{-a}[H^+] + k_{BH^+}[BH^+]) \tau_o'}{1 + (k_a + k_B[B]) \tau_o + (k_{-a}[H^+] + k_{BH^+}[BH^+]) \tau_o'} \qquad (2.15)$$

and

$$\phi'/\phi_o' = \frac{(k_a + k_B[B]) \tau_o}{1 + (k_a + k_B[B]) \tau_o + (k_{-a}[H^+] + k_{BH^+}[BH^+]) \tau_o'} \qquad (2.16)$$

if proton exchange involves the buffer ions B and BH^+ (Eqn. (2.10)) as well as H_2O and H^+ (Eqn. (2.8));

and

$$\phi'/\phi_o' = \frac{1 + (k_{-b}[OH^-] + k_B[B]) \tau_o}{1 + (k_b + k_{BH^+}[BH^+]) \tau_o' + (k_{-b}[OH^-] + k_B[B]) \tau_o} \qquad (2.17)$$

and

$$\phi/\phi_o = \frac{(k_b + k_{BH^+}[BH^+]) \tau_o'}{1 + (k_b + k_{BH^+}[BH^+]) \tau_o' + (k_{-b}[OH^-] + k_B[B]) \tau_o} \qquad (2.18)$$

if proton exchange entails the buffer ions B and BH^+ (Eqn. (2.10)) as well as OH^- and H_2O (Eqn. 2.9)). In Eqns. (2.8) - (2.18) k_a and k_b are the pseudo first order rate constants for excited state deprotonation of HA and protonation of A^-, by water, respectively, k_{-a} and k_{BH^+} are the second order rate constants for excited state protonation of A^- by H^+ and by BH^+,

respectively,. and k_{-b} and k_B are the second order rate constants
for deprotonation of HA by OH^- and B, respectively. τ_o and τ_o'
are the respective lifetimes of HA and A^- in the lowest excited
singlet state in the absence of excited state proton transfer
(i.e. when pH $<<$ pK_a^* and when pH $>>$ pK_a^*, respectively). A
more elegant statement, by Weller (15), of the variations of
ϕ/ϕ_o and ϕ'/ϕ_o' with $[H^+]$, $[OH^-]$, $[B]$ and $[BH^+]$, accounting for
the fact that some HA and A^- molecules fluoresce without enter-
ing into the steady state reaction, is somewhat more accurate
than Eqns. (2.11) - (2.18) which were derived under the assumption that the dispositions of all excited molecules are governed
by steady state kinetics. However, the non-steady state treatment is more complex and for the present discussion Eqns. (2.11)
- (2.18) will suffice.

The actual form of a given fluorimetric titration curve of
a species whose rate of protonation or dissociation, in the excited state, is comparable to its rate of fluorescence, depends
upon whether or not its excited conjugate form also exchanges
protons with the environment at a rate comparable with that of
fluorescence. Some representative fluorimetric titrations are
shown in Fig. (2.4).

In the absence of buffer ions, if the rates of proton exchange in both conjugate excited species are comparable to their
respective rates of fluorescence, the entire fluorimetric titration interval will be a continuous function of $[H^+]$ (or $[OH^-]$)
as described by Eqns. (2.11) - (2.14). The fluorimetric titration curves of HA and A^- will then appear as sigmoidal curves,
as in the equilibrium situation, but the inflection regions of
these curves will be spread over many more pH units than when
proton exchange in the excited state is very fast or very slow,
compared with the rates of fluorescence. As a rule, the faster
the rates of excited state proton exchange relative to those of
fluorescence, the closer to pH = pK_a^* will occur the inflection
point ($\phi/\phi_o = \phi'/\phi_o' = 0.5$) in the fluorimetric titration. The
slower the rates of excited state proton exchange relative to

fluorescence, the closer to pH - pK_a will occur the inflection
point in the fluorimetric titration.

If, in the absence of buffer ions, the rate of reaction of
HA (as in Eqn. (2.8)) or of A^- (as in Eqn. (2.9)), with the
solvent, is comparable in rate to fluorescence of HA or A^-, but
the rate of second order protonation of A^- or deprotonation of
HA is much slower than the rate of fluorescence (i.e. does not
occur during the lifetime of the excited state), Eqns. (2.11) -
(2.14) can be simplified to

$$\left[\frac{\phi}{\phi_o} \right]_{const.} = \frac{1}{1 + k_a \tau_o} \tag{2.19}$$

$$\left[\frac{\phi'}{\phi'_o} \right]_{const.} = \frac{k_a \tau_o}{1 + k_a \tau_o} \tag{2.20}$$

$$\left[\frac{\phi'}{\phi'_o} \right]_{const.} = \frac{1}{1 + k_b \tau'_o} \tag{2.21}$$

and

$$\left[\frac{\phi}{\phi_o} \right]_{const.} = \frac{k_b \tau'_o}{1 + k_b \tau'_o} \tag{2.22}$$

where the const. notation refers to the independence of ϕ/ϕ_o
and ϕ'/ϕ'_o on $[H^+]$ or $[OH^+]$ in Eqns. (2.19) - (2.22). In this
case plotting ϕ/ϕ_o or ϕ'/ϕ'_o vs $[H^+]$ or $[OH^-]$ yields a flat
horizontal line for each species. The independences of ϕ/ϕ_o
and ϕ'/ϕ'_o on $[H^+]$ or $[OH^-]$ may be due to small values of k_{-a}
relative to $1/\tau'_o$, to small values of k_{-b} relative to $1/\tau_o$, or
of very small values of $[H^+]$ or $[OH^-]$. The latter circumstance
is independent of molecular properties and is universal in the
mid pH region. Thus in the mid pH region, the reactions A^- +
$H^+ \rightarrow HA$ and $BH^+ + OH^- \rightarrow B + H_2O$ do not, as a rule, occur during
the lifetime of the excited state and the invariance of ϕ/ϕ_o
and ϕ'/ϕ'_o with $[H^+]$ or $[OH^-]$ is a common phenomenon. At pH

where $k_{-a}\tau_o'[H^+]$ and $k_{-b}\tau_o[OH^-]$ become comparable to $k_a\tau_o$ and $k_b\tau_o'$, respectively, (i.e. near pH = pK_a^*), ϕ/ϕ_o and ϕ'/ϕ_o' vary continuously with $[H^+]$ or $[OH^-]$ according to Eqns. (2.11) – (2.14).

If, in the absence of buffer ions, the rate of reaction of A^- with H^+ or of HA with OH^- is comparable in rate to the fluorescence of A^- or HA, but the rate of dissociation of HA or protonation of A^- by the solvent is much slower than fluorescence (i.e. $k_a \ll 1/\tau_o$ and $k_b \ll 1/\tau_o'$ so that HA $\rightarrow$ H^+ + A^- and H_2O + $A^- \rightarrow$ HA + OH^- do not occur), the fluorimetric titration obtained by exciting HA is not exactly reciprocal to that obtained by exciting A^-. In Eqns. (2.11) and (2.12) the negligibility of $k_a\tau_o[H^+]$ (when HA is excited) leads to

$$\phi/\phi_o = 1$$

and

$$\phi'/\phi_o' = 0$$

i.e. ϕ/ϕ_o and ϕ'/ϕ_o' are independent of pH, if HA is excited, so that the fluorimetric titration curves follow ground state titration characteristics (inflection occurs at pH = pK_a). However, if A^- is excited (pH > pK_a), A^- is a stronger base in the excited state than in the ground state and the pH is, say less than 3, protonation of A^- in the excited state is possible. We then have

$$\frac{\phi'}{\phi_o'} = \frac{1}{1 + k_{-a}\tau_o'[H^+]} \qquad (2.23)$$

and

$$\frac{\phi}{\phi_o} = \frac{k_{-a}\tau_o'[H^+]}{1 + k_{-a}\tau_o'[H^+]} \qquad (2.24)$$

Similarly, in Eqns. (2.13) - (2.14), when A^- is excited we have, $\phi'/\phi_o' = 1$ and $\phi/\phi_o = 0$, so that excitation of A^- yields ground state titration characteristics. Excitation of HA (pH < pK_a) in basic solution, however, where HA is a stronger acid in the excited state than in the ground state and pH > 11 (i.e. $[OH^-]$ is appreciable) results in the likelihood of deprotonation of HA in the excited state, as reflected in the fluorescence of HA and A^- according to

$$\frac{\phi}{\phi_o} = \frac{1}{1 + k_{-b}\tau_o[OH^-]} \qquad (2.25)$$

and

$$\frac{\phi'}{\phi_o'} = \frac{k_{-b}\tau_o[OH^-]}{1 + k_{-b}\tau_o[OH^-]} \qquad (2.26)$$

Eqns. (2.22) - (2.26) are continuous functions of $[H^+]$ or $[OH^-]$ yielding sigmoidal fluorimetric titration curves whose inflection regions are narrow enough for them often to be confused with fluorimetric titrations corresponding to the attainment of equilibrium in the excited state. They may be distinguished from the latter by prediction of the pK_a^* from the spectral shifts accompanying protonation or dissociation. This will be discussed shortly.

Buffer ions are usually employed, for pH control, in solutions having pH between 3 and 11. Moreover, buffer systems are normally chosen in such a way that both conjugate buffer species B and BH^+ are appreciably present in solution (i.e. pH $\sim pK_B$, where pK_B is the reciprocal log of the dissociation constant, K_B, of the buffer ionization equilibrium $BH^+ \rightleftarrows B$ and BH^+). Consideration of these facts which dictate that $[H^+]$ and $[OH^-]$ are very small and [B] and $[BH^+]$ comparable to each other in magnitude, simplifies Eqns. (2.15) - (2.18) to

$$\frac{\phi}{\phi_o} = \frac{1 + k_{BH^+}[BH^+]\tau_o'}{1 + (k_a + k_B[B])\tau_o + k_{BH^+}[BH^+]\tau_o} \qquad (2.27)$$

$$\frac{\phi'}{\phi_o'} = \frac{(k_a + k_B[B])\tau_o}{1 + (k_a + k_B[B])\tau_o + k_{BH^+}[BH^+]\tau_o'} \qquad (2.28)$$

$$\frac{\phi'}{\phi_o'} = \frac{1 + k_B[B]\tau_o}{1 + (k_b + k_{BH^+}[BH^+])\tau_o' + k_B[B]\tau_o} \qquad (2.29)$$

and

$$\frac{\phi}{\phi_o} = \frac{(k_b + k_{BH^+}[BH^+])\tau_o'}{1 + (k_b + k_{BH^+}[BH^+])\tau_o' + k_B[B]\tau_o} \qquad (2.30)$$

if dissociation of HA and protonation of A^- by water are competitive with dissociation of HA by B and protonation of A^- by BH^+. If the buffer ion concentrations are low (say $< 10^{-3}$M) Eqns. (2.27) - (2.30) are reduced to Eqns. (2.19) - (2.22) indicating that reaction of the excited species with the buffer ions does not appreciably occur. In the event that k_a and k_b are very small Eqns. (2.27) - (2.30) reduce to

$$\frac{\phi}{\phi_o} = \frac{1 + k_{BH^+}[BH^+]\tau_o'}{1 + k_B[B]\tau_o + k_{BH^+}[BH^+]\tau_o'} \qquad (2.31)$$

$$\frac{\phi'}{\phi_o'} = \frac{k_B[B]\tau_o}{1 + k_B[B]\tau_o + k_{BH^+}[BH^+]\tau_o'} \qquad (2.32)$$

$$\frac{\phi'}{\phi_o'} = \frac{1 + k_B[B]\tau_o}{1 + k_{BH^+}[BH^+]\tau_o' + k_B[B]\tau_o} \qquad (2.33)$$

and

$$\frac{\phi}{\phi_o} = \frac{k_{BH^+}[BH^+]\tau_o'}{1 + k_{BH^+}[BH^+]\tau_o' + k_B[B]\tau_o} \qquad (2.34)$$

in which case excited state proton transfer is effected only by the buffer ions. At any given concentration of $[BH^+] + [B]$ the fluorimetric titration curves of HA and A^-, as a function of pH will describe somewhat discontinuous inflection regions over the entire interval where $[BH^+]$ is high enough to effect protonation of A^- and $[B]$ is high enough to effect dissociation of HA. $[B]$ and $[BH^+]$ depend, of course, on pK_B as well as upon the total analytical buffer concentration $[B] + [BH^+]$. However, if the concentrations of buffer ions and the rate constants k_B and k_{BH^+} are high enough, the excited state equilibrium represented by Eqn. (2.10) will occur. Moreover, the constant of this equilibrium (K_{B-HA}^*) is related to the equilibrium constant K_a^* for reactions (2.8) and (2.9) by

$$K_a^* = K_B K_{B-HA}^* \qquad (2.35)$$

which indicates that the condition of excited state equilibrium in Eqn. (2.10) also fulfills the conditions of equilibrium in Eqns. (2.8) and (2.9). This result is reasonable because thermodynamically, the definition of equilibrium is independent of the mechanism of approach to equilibrium. Now, at any values of $[B]$ and $[BH^+]$ high enough to sustain equilibrium, in the lowest excited singlet state, between HA and A^-, if pK_a^* is close to pK_B the variations of ϕ/ϕ_o and ϕ'/ϕ_o' with pH will describe the excited state equilibrium fluorimetric titrations corresponding to Eqns. (2.8) or (2.9), because HA and A^- are the only fluorescent species in the system. The dependence of ϕ/ϕ_o and ϕ'/ϕ_o' on buffer concentration indicates that the analyst should be wary of complications in fluorimetric analysis likely to occur in strongly buffered media. This suggests that a tradeoff between maximum buffer capacity and fluorimetric accuracy may be necessary.

The effects of different buffer systems on excited state prototropism in β-naphthol have been studied by Weller (15,16) and in 1- and 2-anthroic acid by Schulman and Capomacchia (17). These investigators have found that the choice of buffer has a profound effect upon the kinetics of excited state prototropism but that the position of the equilibrium is unaffected by the buffer solutions. The excited anthroic acids and their anions (17) which do not equilibrate in dilute acid or base solutions in water may be driven to equilibrium by high concentrations of buffer salts (e.g. 1M $H_2PO_4^-$ - $HPO_4^=$) which act as intermediate proton donors and acceptors.

Due to the dependence of the relative fluorescence efficiency upon excited state protonation kinetics, it is possible to employ the variations of relative fluorescence efficiencies with solution acidity, to study the rapid kinetics of excited state protolysis without the necessity of employing elegant apparatus. It is conceivable that the dependence of relative fluorescence efficiencies upon the kinetics of excited state protolytic reactions can be utilized to devise kinetic methods of analysis which will not be subject to the errors inherent in interrupting progressing chemical reactions.

As an alternative to the method of fluorimetric titration, pK_a^* values can be evaluated from a thermodynamic cycle (Fig. 2.5) due to Förster (5). From Fig. (2.5) it can be seen that there are two mechanistically different but energetically equivalent ways of converting the ground state acid A, to the excited base B^*. The first pathway consists of the absorption of electronic energy E_A by A, to form A^*, followed by dissociation of A^* to B^* with attendant enthalpy of dissociation ΔH^*. The total energy involved in the latter process is thus E_A + ΔH^*. The second pathway from A to B^* entails dissociation of A to B in the ground state, accompanied by the enthalpy of dissociation ΔH, followed by excitation of B to B^*, the electronic energy E_B being absorbed. This second mechanism requires a total energy of E_B + ΔH. Since both mechanisms are thermodynamically equivalent

$$E_A + \Delta H^* = E_B + \Delta H \qquad (2.36)$$

In Eqn. (2.36) E_A and E_B are, respectively, equal to $Nh\nu_A$ and $Nh\nu_B$ where N is Avogadro's number, h is Planck's constant and ν_A and ν_B are the frequencies of radiation involved in the transitions between A and A^* and B and B^*, respectively. If it is assumed that the entropies of protonation in ground and excited states are identical, then Eqn. (2.36) becomes

$$\Delta G - \Delta G^* = Nh(\nu_A - \nu_B) \qquad (2.37)$$

from which it immediately follows that

$$pK - pK^* = \frac{Nh}{2.303RT}(\nu_A - \nu_B) \qquad (2.38)$$

where R is the universal gas constant and T is the absolute temperature. Finally, if the constants are expressed in c.g.s. units, at $25^\circ C$ we have

$$pK - pK^* = 2.10 \times 10^{-3}(\bar{\nu}_A - \bar{\nu}_B) \qquad (2.39)$$

where $\bar{\nu}_A$ and $\bar{\nu}_B$ are the wavenumbers (cm^{-1}) of the transitions from A to A^* and B to B^*, respectively. Consequently, if the transition energies of acid and conjugate base are known, and the ground state pK_a can be evaluated, the excited state pK_a^* for the corresponding equilibrium can be calculated. The accuracy with which pK_a^* can be calculated from the Förster cycle is usually established by comparison with pK_a^* values obtained from fluorimetric titration and is subject to the validity of the assumption of equal entropies of protonation in ground and electronically excited states and to the way in which $\bar{\nu}_A$ and $\bar{\nu}_B$ are chosen. These subjects are worthy of some discussion. In aromatic structures, in which gross changes in molecular electronic structure do not occur subsequent to excitation, the entropies of protonation in ground and excited states which are

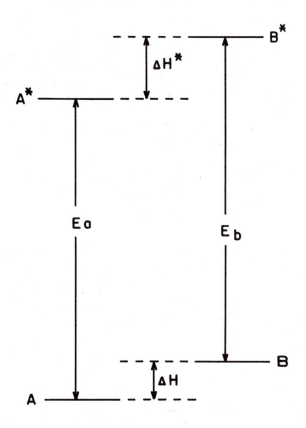

THE FÖRSTER CYCLE

Fig. 2.5. The Förster cycle - A,A*, B and B*, represent the
          ground and excited states of acid and conjugate base.
          E_A and E_B are the spectroscopic energies of the tran-
          sitions A* $\rightleftarrows$ A and B* $\rightleftarrows$ B, respectively, and ΔH and
          ΔH^* are the enthalpies of protonation in the ground
          and electronically excited states, respectively.

predominately related to configurational changes accompanying
protonation or dissociation, are not very different (18). Thus
in most conjugate acid-base pairs there is good agreement between pK_a^* values calculated from the Förster cycle and those
determined by fluorimetric titration. In some cases, however,
protonation or dissociation causes electronic structural changes
in acid or conjugate base, so that the electronic structure of
one may not correspond to the electronic structure of the other
as in 2-hydroxyquinoline (19) which is phenolic in its cation
and anion forms but is a cyclic amide in its uncharged form.
Occasionally, the structure of one species changes dramatically
upon excitation while the structure of its conjugate acid or
base does not. For example, 9-anthroic acid has the carboxyl
group perpendicular to the aromatic ring in the ground state and
coplanar and conjugated with the aromatic ring in the lowest
excited singlet state. In the 9-anthroate anion, the carboxylate
group is perpendicular in ground and excited states (5). Finally,
it is occasionally observed that the ^1L_a and ^1L_b states in acid
and conjugate base are opposite in order, as in the case of 1-
naphthylamine and the 1-naphthylammonium ion (20). In all three
of the preceding situations the differences in electronic structures of conjugate acids and bases cause the pK_a^* values determined from the Förster cycle to be substantially different
from those determined by titrimetry. These discrepancies are
believed to result from the violation of the assumption of equal
entropies in ground and excited singlet states, inherent in the
Förster cycle.

The quantities $\bar{\nu}_A$ and $\bar{\nu}_B$ in Eqn. (2.39) represent the wavenumbers of "pure" electronic transitions in acid and conjugate
base so that $\bar{\nu}_A - \bar{\nu}_B$ should not contain any vibrational energy.
To a first approximation, $\bar{\nu}_A$ and $\bar{\nu}_B$ may be taken from either the
long wavelength absorption bands of acid and conjugate base,
respectively, or (if both members of the conjugate pair are
fluorescent) from the fluorescence bands of acid and conjugate
base, respectively. To avoid inclusion of vibrational energy in
the Förster cycle calculation, $\bar{\nu}_A$ and $\bar{\nu}_B$ should be taken at the

0-0 vibronic bands of absorption or fluorescence of acid and
conjugate base, respectively. However, in most compounds of
interest, the positions of the 0-0 bands are generally not
known, especially at room temperature. Levshin (21) has shown
that if the spacings between vibrational sublevels in ground and
electronically excited states of the same molecule are identical,
an approximate mirror image relationship exists between the
fluorescence spectrum and the long wavelength absorption band.
If a mirror image relationship exists between absorption and
fluorescence, the band maxima of fluorescence and absorption are
equally displaced, in energy, from the 0-0 band (the 0-0 band
ideally represents the long wavelength limit of absorption and
the short wavelength limit of fluorescence). Thus by averaging
the energies (or wavenumbers) of the fluorescence and long wavelength absorption maxima, the position of the 0-0 band could be
approximated. If both members of a conjugate pair are fluorescent, it is then possible to calculate the wavenumbers of the
0-0 band for each and apply the Förster cycle. Probably more
pK_a^* values in the literature have been obtained by the latter
means than by any other.

The calculation of pK_a^* values from the Förster cycle
employing the average of absorption and fluorescence maxima is
applicable only to those conjugate pairs in which both members
are fluorescent. However, if the assumption is made that the
vibrational spacings in ground and excited states of both members of a conjugate pair are identical, then the absorption
band maxima of acid and conjugate base will be equally displaced
from the respective 0-0 bands (22). In this case the absorption
maxima may be used in the Förster cycle to calculate pK_a^*, the
vibrational discrepancies dropping out in the difference $\bar{\nu}_A -
\bar{\nu}_B$. It should be noted that in this case it is not necessary
for either member of a conjugate pair to be fluorescent in order
to calculate a pK_a^*. In the event that neither number of the
conjugate pair fluoresces, pK_a^* is meaningless from the kinetic
point of view since lack of fluorescence implies that the excited molecule converts to the ground state vibrationally, in a

time too short for excited state prototropism to occur. However,
the thermodynamic significance of such a pK_a^* is presumably the
same as if excited state prototropism did occur.

Calculation of excited state pK_a^* values from the Förster
cycle employing absorption data alone entails another source of
error apart from vibrational considerations. In water, most
molecules experience substantial stabilization, subsequent to
excitation, as a result of solvent reorientation about the excited dipole. If solvent relaxation occurs subsequent to excitation and it occurs unequally for acid and conjugate base,
pK_a^* values determined by fluorimetric titration or from shifts
of the fluorescence spectrum alone, in conjunction with the
Förster cycle, will reflect the effect of solvent relaxation,
since the latter process affects the energies of the species
involved in the excited state equilibrium and occurs during the
lifetime of the excited state. Because absorption occurs in a
time which is short compared with the time required for solvent
relaxation, the absorption spectra will not reflect the effect
of excited state solvent relaxation and pK_a^* values calculated
from absorption data alone will be in error. In the latter
regard, it has been found that solvent relaxation errors may
also be present in pK_a^* values determined from the Förster cycle
employing the average of absorption and fluorescence maxima.
The pK_a^* value of doubly protonated o-phenanthroline determined
by fluorimetric titration, indicates that doubly protonated
phenanthroline is a stronger acid in the lowest excited singlet
state than in the ground state (11). Calculations of pK_a^* from
the Förster cycle, for the latter acid, employing fluorescence
shifts alone, were in qualitative agreement with the fluorimetric titration. However, absorption data alone and the averaging
of absorption and fluorescence shifts all suggested that o-
phenanthroline dication was a weaker acid in the excited state.
These results also serve to illustrate that in some cases the
effect of solvent relaxation in the excited state may actually
reverse the effect of the electronic shift, occurring upon excitation, in determining the acidity of the excited molecule.

However, in many cases pK_a^* values calculated from absorption
data alone may serve as useful approximate indices of excited
state chemistry, especially in cases where the fluorescence from
one or both members of a conjugate pair is non-existent or where
prototropism in the excited state is so slow that an accurate
treatment is not possible.

Application of the Förster cycle to the study of excited
state prototropism contains several potential pitfalls of which
the investigator should be aware. The pK_a and pK_a^* values in
Eqn. (2.15) correspond to the same equilibria (i.e. dissociation
from the same functional group) in ground and excited states.
In polyfunctional molecules, dissociation frequently occurs from
different groups, in ground and excited states, as a result of
the difference in charge distributions in those states. For
example, consider a solution of 7-quinolinol in acid solution
(23,24). As the pH is raised, the 7-quinolinol cation, in the
ground state, loses a proton from the ring nitrogen to form the
neutral molecule. In the ground state, a zwitterionic species
is present in neutral solutions only as a small fraction of the
concentration of the neutral species. As the solution is made
basic, the neutral species dissociates from the phenolic group
to form the anion. In the lowest excited singlet state, however, the phenolic group is more acidic and the ring nitrogen
more basic, so that in acid solution dissociation occurs from
the phenolic group to form the zwitterion while in basic solutions, the zwitterion dissociates from the ring nitrogen to form
the anion. Clearly, the ground state and excited state pK_a
values will not correspond to the same equilibria; nor will
absorption and fluorescence spectra correspond to the same
species. The proper solution to this problem lies in preparing
methylated derivatives of the molecule and evaluation of the
ground state titrimetric and spectral properties of the N-
methylated and O-methylated derivatives. In basic or neutral
solutions the N-methylated and O-methylated derivatives are
models of the zwitterionic and neutral species, respectively.
Comparison of the fluorescences of the methylated derivatives

with those of the unmethylated 7-quinolinol allows interpretation
of the observed equilibria and proper employment of the Förster
cycle. The $\bar{\nu}_A$ and $\bar{\nu}_B$ terms in the Förster cycle are presumed to
belong to transitions from the ground states of acid and conju-
gate base, respectively, to corresponding electronically excited
states. The term "corresponding" as employed here, means that
the electronic configurations and angular momentum quantum numbers of the excited states of acid and conjugate bases are
identical (i.e. the presence or absence of the dissociating
proton is regarded as producing a weak perturbation). Occasionally, it may be found that the excited states in acid and conjugate base do not correspond because protonation or dissociation
either changes the ordering of states or the molecular structure.
Thus dissociation in the excited state entails an electronic
transition as well as a chemical reaction. This phenomenon will
usually introduce a large entropy (associated with configurational change) error into the Förster cycle calculation. For
example, the molecule 3,4-benzcinnoline (25) fluoresces from an
$n-\pi^*$ lowest excited singlet state. As acid is added to a solution of 3,4-benzcinnoline the fluorescence is quenched. Fluorimetric determination of pK_a^* indicates that the molecule is a
stronger acid in the excited state than in the ground state as
is expected for a molecule in which the optical electron is
nonbonding. However, the shift of the long wavelength $\pi-\pi^*$
band of 3,4-benzcinnoline upon protonation indicates that the
molecule is a weaker acid in the excited state. Moreover, in
the conjugate acid the lowest excited singlet state is $\pi-\pi^*$.
Thus if the natures of fluorescing states or the presence of
hidden absorption bands are not reckoned with, correlations of
Förster cycle data with fluorimetric titration data may be
invalid.

    So far, it has been shown that Förster cycle calculations of
pK_a^* values are, at best, approximate. Those pK_a values calculated from fluorescence maxima alone may contain errors due to
the inclusion of vibrational energy and neglect of ground state
solvent relaxation. Those calculated from the averaging of

absorption and fluorescence maxima may contain errors due to
unequal solvent relaxation of acid and conjugate base subsequent
to excitation while those calculated from absorption maxima alone
may contain errors due to the inclusion of vibrational energy
and neglect of excited state solvent relaxation. However, useful
qualitative information regarding the chemistry of the lowest
excited singlet state can be obtained from Förster cycle calculations. Moreover, it is the very errors inherent in the Förster
cycle calculations that yield information about the excited state
processes causing these errors when the Förster cycle data are
compared with pK_a^* values determined by fluorimetric titration.

Some correlations between electronic and chemical structure
and pK_a^* values are possible from the studies currently in the
literature. Some of the important relationships between structure and excited state prototropic reactivity will now be sum-
marized.

Phenolic groups, aromatic amines and sulfhydryl groups
generally tend to become more acidic in the lowest excited singlet state relative to the ground state (16). Carbonyl groups,
carboxyl groups and nitrogen atoms in six-numbered heterocyclic
rings tend to become more basic in the lowest excited singlet
state than in the ground state (16). Exceptions to the latter
rule, however, are to be found in the cases of 3,4-benzcinnoline
(25) where the nitrogen atom becomes more acidic in the excited
n-π* state and in doubly protonated o-phenanthroline where solvent relaxation opposes charge-transfer in the $^1\pi-\pi^*$ excited
state (11). Five-membered nitrogen heterocycles (e.g. indole
and carbazole) have been shown to behave more as secondary
arylamines, becoming less basic and more acidic in the excited
state at the nitrogen atom, than as six-membered nitrogen heterocycles (26). In indole and carbazole protonation occurs at
carbon rather than at nitrogen in both ground and lowest excited singlet states.

Valence shell expansion of substituent atoms with vacant
d orbitals ($d_\pi - p_\pi$ bonding) has been observed in the ground and

excited states of several organic sulfur compounds. The enhancement of the acidity of excited phenols substituted with groups
containing sulfur atoms bearing a formal positive charge is due
to valence shell expansion to sulfur in the lowest excited singlet state (27). In this process the vacant 3d orbitals of sulfur
accept charge from the aromatic system and thereby exert an acidity strenghtening effect. It was shown, by fluorimetric titration, that the anomalously low pK_a^* value for the lowest excited
singlet state equilibrium between the cation and zwitterion
derived from 8-mercaptoquinoline, and the anomalously short
emission wavelength of the 8-mercaptoquinoline cation relative
to that of 8-hydroxyquinoline, could be explained by valence
shell expansion of the sulfur in the protonated 8-mercaptoquino-
line in the lowest excited singlet state. Moreover, since the
shifts in the absorption spectra of 8-mercaptoquinoline upon
protonation are similar to the shifts in the absorption spectra
of 8-hydroxyquinoline upon protonation, valence shell expansion
of sulfur must occur subsequent to excitation suggesting that
the sulfur atom of the 8-mercaptoquinoline cation acquires a
formal positive charge only after excitation (28).

In conclusion, it can be seen that due to the rapidity
with which excited state protolytic processes occur in fluid
solutions, comparisons of the pH dependences of absorption
spectra with those of fluorescence spectra can yield valuable
information concerning the photochemistry of the lowest excited
singlet state.

The pH dependence of phosphorescence. Because of the long
lifetimes of molecules in electronically excited triplet states,
nonradiative vibrational and collisional processes compete effectively with phosphorescence for deactivation of the lowest
excited triplet state. It is this very fact that makes the
photochemistry of the triplet state so much more diverse than
that of the lowest excited singlet state. However, effective
competition of nonradiative deactivational processes with phosphorescence makes phosphorescence in fluid solutions extremely

rare. As a result, phosphorimetry is usually practiced in rigid
solutions at liquid nitrogen temperatures. Thus diffusion limited processes (e.g. proton transfer) do not affect the phosphorescence of rigid solutions. However, the arguments that apply to
fluorimetry in rigid solutions or in cases where excited state
prototropism is slow, in fluid solutions, compared with fluorescence, also apply to analytical phosphorimetry. Although it is
the ground state thermodynamics of the solution, prior to freezing, that determines which conjugate species phosphoresces, the
quantum yields of phosphorescence may be very different for acid
and conjugate base and a study of the pH dependence of phosphorescence is therefore desirable. For example, the quantum
yield of phosphorescence of the p-nitrophenolate anion is some
10^3 times greater than that of the neutral p-nitrophenolate
molecule (13). It would, therefore, be most desirable to carry
out phosphorimetric analysis for p-nitrophenol in basic solution.
Quantitative phosphorimetry is generally limited to those solvents which form clear glasses or snows upon freezing. Of these
solvents, only ethanol and aqueous alcohol have the required
amphiprotic properties necessary to carry out a substantial pH
variation (13). Thus, phosphorimetry in these solvents may
have some inherent advantages over phosphorimetry in solvents
of lower dielectric strength and hydrogen bonding capabilities.

In spite of the relatively limited dependence of phosphorescence upon pH in rigid solutions, some useful information
concerning the chemistry of the lowest excited triplet state is
to be had from the study of the shifts of phosphorescence
occurring as a result of protonation and dissociation. The
Förster cycle (Eqn. 2.15) may be applied to pH dependent shifts
of phosphorescence spectra in the same way it is applied to
shifts of absorption or fluorescence spectra. Thus if the pK_a
for a particular prototropic equilibrium is known and $\bar{v}_A$ and
$\bar{v}_B$ are taken to be the 0-0 bands (the low frequency vibrational
features) of the phosphorescence spectra of acid and conjugate
base, respectively, the dissociation constant (pK_T^*) for the
acid in the lowest excited triplet state can be calculated.

Because phosphorescence is measured in rigid solutions where
proton transfer cannot occur subsequent to excitation, the pK_a^*
will always correspond to the same equilibrium to which the
ground state pK_a corresponds. At low temperatures the vibrational structure of electronic spectra, often not apparent in
fluid solutions at room temperature is usually more obvious.
The 0-0 band of phosphorescence can usually be identified or
at least approximated from the position of the shortest wavelength vibronic maximum in the phosphorescence spectrum (6).
Thus errors in the estimation of pK_T^* due to the inclusion of
vibrational energy are virtually non-existent. This is a
fortunate circumstance since the averaging of absorption and
phosphorescence spectra in this case would require the measurement of singlet-triplet absorption spectra which are, in most
cases, impossible to obtain due to the extremely low probability of direct singlet-triplet electronic processes.

An alternative but more complex method for the determination
of pK_T^* values is available in the form of flash photolysis, to
produce an appreciable number of triplet molecules, followed by
rapid determination of the pH dependence of the triplet-triplet
absorption spectrum. The results obtained from the flash photolysis method for several aromatic phenols, carboxylic acids,
amines and heterocycles have been in excellent agreement with
those obtained from phosphorescence shifts employed in conjunction with the Förster cycle (6). Because the triplet-triplet absorption spectra, employed with flash photolysis, are taken
on fluid solutions containing triplet molecules which are fully
relaxed with respect to both vibrational manifold and solvent
cage, the agreement between the phosphorescence and triplet-
triplet absorption data indicate that vibrational errors are
indeed small in the phosphorescence shift approach and that solvent relaxation is not important as a factor in determining the
value of pK_T^*. The latter result is a reasonable one because
the high degree of electronic repulsion in the triplet state
results in smaller electronic polarization differences and thus
smaller solvent effects than in the lowest excited singlet state

relative to the ground state. As a result of the latter circumstances, differences between ground and triplet state dissociation constants tend to be much smaller than differences between
ground and lowest excited singlet state pK_a values. Usually,
pK_T^* values will be intermediate between pK_a and pK_a^* values.
However, occasionally the lowest excited triplet state may have
a different direction of polarization than the lowest excited
singlet state or solvent relaxation may be important in the
singlet state in which case the relative values of pK_a^* and pK_T^*
will not be easily predictable. For example, the lowest excited
singlet state of o-phenanthroline and its various protonated
forms is of the ^1L_b type. However, the lowest excited triplet
state of o-phenanthroline is of the ^3L_a type. These transitions
have their moment vectors directed perpendicular to one another.
In the lowest excited triplet state the dication of o-phenanthroline becomes more acidic while the monocation of o-phenanthroline becomes less acidic than in the ground state (30). This
is in agreement with the direction of the $^3L_a \rightarrow {}^1A$ transition
moment (11). However, in the lowest excited singlet state the
direction of the $^1L_b \rightarrow {}^1A$ transition moment suggests that both
the dication and monocation should be weaker acids than in the
ground state. Actually, due to the prevalence of a large solvent relaxation phenomenon in the equilibrium between the
dication and monocation, the dication does become more acidic
in the lowest excited singlet state with pK_a^* being almost the
same as pK_T^*.

Phototautomerism

     In the previous section it was stated that electron with-
drawing groups become more basic or less acidic in the lowest
excited singlet and triplet states and that the increases in
basicity were associated with spectral shifts to lower frequency
upon protonation while the decreases in acidity were associated
with spectral shifts to higher frequency upon dissociation.
Moreover, electron donating groups become more acidic or less
basic in the lowest excited singlet and triplet states and

increases in acidity were associated with spectral shifts to
lower frequency upon dissociation while decreases in basicity
were associated with spectral shifts to higher frequency with
protonation.   These generalizations are valid for polyfunctional
as well as monofunctional aromatic molecules.   In certain polyfunctional molecules, especially those containing at least one
basic electron accepting group and at least one acidic electron
donating group, the acceptor group may become so basic and the
donor group so acidic, subsequent to excitation, that a proton
will be simultaneously lost by the donor group and gained by
the acceptor group.   This amounts to an isomerization or tautomerization in the excited state (phototautomerism), there being
no net gain or loss of protons by the molecule having undergone
phototautomerization.   If a polyfunctional molecule undergoes
phototautomerization, no indication of this process is observed
in the electronic absorption spectrum because tautomerization
occurs subsequent to the absorption process.   However, if
phototautomerism occurs during the lifetime of the lowest excited
singlet state it will be very much in evidence in the fluorescence spectrum.   If phototautomerism involves protonation of an
electron acceptor group and dissociation of an electron donor
group the shift to lower frequency (relative to the unphoto-
tautomerized molecule) produced by phototautomerism will be
very nearly the sum of the frequency shifts produced by protonation and dissociation individually.   Thus, an anomalously
large frequency difference between the long wavelength absorption maximum and the fluorescence maximum is often an
indication that phototautomerism has taken place in the lowest
excited singlet state.   Moreover, the order of sequential protonations and dissociations of the phototautomer are opposite
to the order of protonations and dissociations of the "normal"
molecule (e.g. the acceptor group dissociates first in the
ground state but not in the excited state).   Thus opposite
sequences of spectral shifts of absorption and fluorescence
shifts with changing pH (e.g. a shift to lower frequency
followed by a shift to higher frequency of the absorption spectrum and a shift to higher frequency followed by a shift to

lower frequency of the fluorescence spectrum with decreasing
pH) are also often indicative of phototautomerism in the lowest
excited singlet state.

Two types of phototautomerism, intramolecular and biprotonic, may be distinguished. Intramolecular phototautomerism
entails the direct transfer of a proton from the electron donor
group to the electron acceptor group. This process has been
observed only in polyfunctional molecules in which both groups
are situated ortho to one another. Usually, there is a fairly
strong intramolecular hydrogen bond between a proton of the
donor group and the acceptor group, in the ground state. His-
torically, the first example of phototautomerism observed was
the intramolecular phototautomerism of salicylic acid (o-hy-
droxybenzoic acid). Weller (31) observed that the fluorescence
of salicylic acid in polar solvents occurred at very low frequencies. In hydrocarbon solvents salicylic acid exhibited two
fluorescence bands. One was at low frequency and was similar to
the emission observed in ethanol. The other was at higher frequency, formed a mirror image with the low frequency absorption
band of salicylic acid, and was similar in position to the
fluorescence of o-methoxybenzoic acid, a molecule similar in
electronic structure to salicylic acid but lacking a phenolic
proton which could be transferred to the carboxyl group. From
these pieces of evidence Weller concluded that the absorption
and high frequency fluorescence of salicylic acid in hydrocarbon
solvents was due to the neutral molecule (N) but that the low
frequency fluorescence was due to the zwitterion (Z) formed as
a result of phototautomerism in the lowest excited singlet state.

(N)   (Z)

Because phototautomerization of salicylic acid occurred in
hydrocarbon solvents in which the solvent could not assist in
proton transfer, the process had to be (at least in hydrocarbon
solvents) intramolecular. Subsequent studies of the pH dependence of the fluorescence of salicylic acid (32) showed the
phototautomerism to be independent of pH indicating that the
process was intramolecular even in water. In the studies of
the pH dependence of the fluorescence of salicylic acid it
was also shown that although the titration characteristics
paralleled those of the ground state of the neutral molecule,
the emission always occurred from the phototautomer. Thus
even though the protonation and dissociation phenomena of
salicylic acid, in the excited state, were too slow to achieve
equilibrium, the intramolecular proton transfer process was
essentially complete by the time fluorescence occurred.

Phototautomerism has been observed in molecules in which
the donor and acceptor groups are widely separated. In these
cases, phototautomerism must be biprotonic, taking place in
two steps; protonation of the acceptor group and dissociation
of the donor group (or vice versa). In this case, the role of
solvent is critical because the solvent is responsible for
transporting protons to and from prototropic sites in the molecule. Because two distinct prototropic processes in widely
separated groups are involved, the biprotonic process is subject
to the same kinetic considerations as the attainment of prototropic equilibrium during the lifetime of the excited state and
as a result biprotonic phototautomerism will usually not occur
in molecules with very short lived excited states or slow protonation and dissociation kinetics.

4-Methyl-7-hydroxycoumarin (β-methylumbelliferone), a
molecule often employed as a fluorescent indicator in kinetic
studies of hydrolytic enzyme reactions, is an example of a molecule which displays biprotonic phototautomerism in the lowest
excited singlet state (33). The low frequency absorption
maxima of 4-methyl-7-hydroxycoumarin and 4-methyl-7-methoxy-

coumarin lie at 32,000 cm^{-1} and 31,000 cm^{-1}, respectively, at
pH 1. However, at pH 1 the fluorescence maximum of the 7-methoxy
compound lies at 25,900 cm^{-1}, while that of the 7-hydroxy compound lies at 21,000 cm^{-1}, suggesting that in the ground state,
the uncharged molecule is the neutral molecule (N) while in the
excited state it is the zwitterion (Z).

(N)   (Z)

This conclusion is substantiated by the fact that upon going
from pH 10 to pH 1, the absorption spectrum shifts to higher
frequency as the phenolate group of the anion is protonated
while the fluorescence spectrum shifts to lower frequency, presumably as the result of protonation of the carbonyl group in
the excited state. Moreover, in chloroform solution, where
intermolecular proton transfer is impossible, the fluorescence
maximum of the 7-hydroxy compound lies at 26,300 cm^{-1}, close
to the emission maximum of the 7-methoxy derivative in which
phototautomerism is impossible.

Phototautomerism is not restricted to proton transfer
between electron acceptor and electron donor functional groups.
In some instances phototautomerism has been observed in bi-
functional molecules in which both functional groups are of
the acceptor type. In some quinoline carboxylic acids in mod-
erately concentrated acid solutions, the singly protonated
cation in the ground state contains a neutral carboxylic acid
group and a protonated nitrogen atom. However, in the lowest
excited singlet state, the carboxyl group is more basic than

the ring nitrogen atom so that the carboxyl group becomes protonated and the ring nitrogen atom deprotonated. For example,
in cinchoninic acid (quinoline-4-carboxylic acid), the fluorimetric titration behavior in acid media suggests that the photo-

tautomerization takes place (34).

Phototautomerization in the lowest excited triplet state
has never been directly observed. This is a consequence of the
fact that phosphorescence, the spectroscopic indicator of triplet state reactivity, is usually observed only in rigid media,
so that proton mobility is virtually eliminated. However, the
triplet state dissociation constants of some nitro substituted
8-hydroxyquinolines, determined from phosphorescence shifts
produced by static protonation and dissociation in conjunction
with the Förster cycle, suggest that triplet state phototautomerism is likely in these compounds (35).

8-Hydroxyquinolines dissociation in the ground state
according to the sequence

However, in the lowest excited singlet state the ring nitrogen
atom becomes much more basic and the phenolic group much more
acidic, with the result that the excited zwitterion (Z) is the
stable uncharged species in the lowest excited singlet state
rather than the neutral molecule (N) as in the ground state.
Thus, the dissociation sequence for 8-hydroxyquinoline and its
simple derivatives, in the lowest excited singlet state, as
observed directly through fluorescence spectroscopy is

(C)   (Z)   (A)

In the lowest triplet states of 5-nitro-8-hydroxyquinoline and
7-nitro-8-hydroxyquinoline it is anticipated that the ring
nitrogen atom will be more basic than in the ground state (but
not as basic as in the lowest excited singlet state) and that
the phenolic group will be more acidic than in the ground state
(but not as acidic as in the lowest excited singlet state).
Thus, phototautomerism is theoretically possible. The triplet
state dissociation constants for the equilibria between the
cations and neutral molecules as (pK_{CN}^{T}) for the 5-nitro and 7-
nitro derivatives of 8-hydroxyquinoline are 5.3 and 6.3, respectively. The values of pK_{NA}^{T} for the triplet state equilibria
between the neutral molecules and anions derived from 5-nitro-
8-hydroxyquinoline and 7-nitro-8-hydroxyquinoline are 4.1 and
6.4, respectively. This rather unusual order of dissociation
constants dictates that the neutral molecules in the triplet
state are stronger or comparable in acidity to the cations.
Consequently, it is to be expected that in the cations, where
the positive charge on the ring nitrogen atoms makes the phenolic groups even stronger acids, dissociation in fluid solutions, in the lowest triplet state will occur from the phenolic

group rather than from the ring nitrogen atoms with the result
that the stable uncharged forms of the 5- and 7-nitro derivatives will be the zwitterions. Thus, it can be predicted that
excitation of 5-nitro-8-hydroxyquinoline and 7-nitro-8-hydroxy-
quinoline to the lowest triplet state, in solution conditions
where the neutral molecules are predominant in the ground state
will result in phototautomerism to the zwitterions in the lowest
triplet state.

It should now be apparent that fluorescence and phosphorescence spectroscopy are powerful tools for the study of the
structure and reactivity of electronically excited molecules.

Quenching of Molecular Luminescence

Fluorescence or phosphorescence intensities may be quenched
(diminished or even eliminated) due to the deactivation of the
excited states responsible for luminescence by interaction of
either the ground or excited states of the luminescing species
with other species in solution. The mechanisms of luminescence
quenching are not completely understood but reversible electron
transfer appears to be involved in quenching by species with
low ionization potentials or electron affinities (e.g. molecular oxygen, transition metal ions and other paramagnetic or
highly conjugated species) (16). On the other hand, evidence
is being accumulated that quenching by interaction with the
solvent, or by interaction with hydrogen bonding molecules pro-
ceeds by a vibrational mechanism; the interaction between the
luminescing molecule and the solvent providing vibrational
coupling which favors efficient internal conversion.

Quenching processes may be divided into two broad categories.
In dynamic or diffusional quenching, interaction between the
quencher and the potential luminescer takes place during the lifetime of the excited state. As a result, the efficiency of dynamic quenching is limited by the lifetime of the excited state of
the potential luminescer and the concentration of the quenching

species. Complexation between quencher and excited molecule
results in excited state electron transfer, forming a nonfluorescent oxidized or reduced species, or on vibrational coupling
of the excited molecule with the solvent. Internal conversion
then returns a fraction or all of the excited molecules, radiationlessly, to the ground state resulting in a diminished
quantum yield of fluorescence or phosphorescence. The quenching
of the blue-green fluorescence of 1-naphththylamine in alkaline
aqueous solutions (37), in which no changes in absorption spectra
are observed, is an example of dynamic quenching. Static quenching is characterized by complexation in the ground state between
the quenching species and the molecule which, when alone excited,
should eventually become a potential fluorescing or phosphorescing species. The complex is generally not capable of luminescing
and although it may dissociate in the excited state to release
the luminescing species, this phenomenon is only partially effective in producing luminescers because the dissociation of the
excited complex may be slow, so that radiationless conversion to
the ground state may be much more efficient than photodissocia-
tion. As a result, the ground state reaction diminishes the
intensity of fluorescence or phosphorescence of the potentially
luminescing species. The quenching of the fluorescence of o-
phenanthroline by complexation with iron (II) is an example of
static quenching.

It should be noted that in dynamic quenching, the quantum
yields of fluorescence and phosphorescence are governed by the
kinetics of photoreaction. However, in static quenching, they
are generally governed exclusively by the strength (thermodynamics) of ground state complexation. Thus conversion of an
excited fluorescing acid or base to its nonfluorescent conjugate
by dissociation or protonation represents static quenching by
H_2O, OH^-, or H^+, if the quenching is centered at pH = pK_a and
dynamic quenching if it is centered at pH $\neq$ pK_a.

The dependence of the relative quantum yield of fluorescence (ϕ/ϕ_o) upon quencher concentration [Q] may be derived from

steady state kinetics and is where τ^{o} is the mean lifetime of

$$\phi/\phi_{o} = \frac{1 - \gamma\alpha}{1 + \gamma k_{D}\tau^{o}[Q]} \qquad (2.40)$$

the excited state in the absence of quenching (i.e. when [Q] = 0), k_{D} is the molecular probability (or on a macroscopic scale, the rate constant) of complexation of the potential fluorescing species by the quencher Q in the excited state, γ is the probability that the complex will be deactivated rather than dissociate back to the potential fluorescer and α is the fraction of complex formed in the ground state that is excited by direct absorption. If complexation in the ground state is negligible, or if excitation is effected at a wavelength where the complex does not absorb, $\alpha \rightarrow 0$ and Eqn. (2.16) is reduced to

$$\phi/\phi_{o} = \frac{1}{1 + k_{Q}[Q]} \qquad (2.41)$$

where $k_{Q} = \gamma k_{D}\tau^{o}$. Eqn. (2.41) is the well known Stern-Volmer equation, which relates the quantum yield of fluorescence to the rate of dynamic quenching and the quencher concentration. Because k_{D} is very specific for quenching species, the sensitivity and selectivity inherent to fluorescence spectroscopy can often be extended to nonfluorescent molecules by making use of their specific abilities to quench the fluorescences of intensely emitting species.

Because phosphorescence is almost exclusively studied in rigid media where diffusional processes do not occur to a measurable extent, the observation of dynamic quenching in dilute solutions, is limited essentially to fluorescence. However, static quenching may be observed for either fluorescence or phosphorescence.

Coordination by Metal Ions

The coordination of fluorescing or phosphorescing aromatic ligands by metal ions is actually an acid-base reaction, with the metal ion acting as a Lewis acid (electron pair acceptor) and the ligand acting as a Lewis base (electron pair donor). In this regard, the coordination of ligands by metal ions is analogous to the protonation of the ligand, in which case, the hydrogen ion functions as the Lewis acid. As a result, it might be expected that many of the changes of the electronic spectra of the ligands, produced by metal ion coordination, will be analogous to the pH and Hammett acidity dependences of ligand spectra described earlier in this chapter. In many cases this is true. However, the parallelisms between electronic spectral changes in the ligand produced by protonation and those produced by coordination with metal ions, especially transition metal ions are, at best, approximate and often there are spectral phenomena observed with metal ion coordination that have no equivalent in coordination with hydrogen ion (and vice versa).

The coordination of aromatic ligands by nontransition metal ions (e.g. Zn(II), Cd(II), Al(III), Ga(III)) has the effect of producing positive polarization at the coordination sites on the ligand and the spectral shifts that are produced by coordination with these ions are similar to the shifts produced by protonating the ligand at the coordination sites. Thus 8-hydroxyquinoline which is colorless and fluoresces blue in ethanolic solutions becomes yellow and fluoresces green when either hydrogen ion or a nontransition metal ion (37) is allowed to react with it in ethanolic solutions. The actual low frequency absorption maxima and fluorescence maxima of the 8-quinolinolium cation and the 8-quinolinolate-nontransition metal complexes differ slightly, as a result of the differences in polarizing abilities between the proton and the various nontransition metal ions, but the absorption maxima are all in the vicinity of 2.7×10^4 cm^{-1} and the fluorescence of the protonated ligand and of the ligand complexed with the various nontransition metal ions are substantially different from one another, the quantum yields of

fluorescence decreasing with increasing atomic number of the
metal ion, as a result of spin orbital coupling which populates
the lowest triplet state at the expense of the lowest excited
singlet state. Moreover, the fluorescence of the 8-hydroxy-
quinolinium cation varies with Hammett acidity in a region of
acidity where the absorption spectrum of this species is in-
variant. This is the result of establishment of prototropic
equilibrium in the lowest excited singlet state between the
cation and zwitterion, derived from 8-hydroxyquinoline, a
situation which can only occur if the rates of complexation
(protonation) and dissociation are rapid compared with the lifetime of the excited state. However, metal ion complexation and
dissociation kinetics are always much slower than protonation
and protolytic dissociation kinetics, the fastest of the former
processes taking about 10^{-5} sec. As a result, when 8-hydroxy-
quinoline is bound by Mg(II), Ba(II), or Al(III) the red shift
of the fluorescence spectrum occurs simultaneously with the red
shift of the absorption spectrum. Thus, in the kinetic aspects
of complexation and the intensity factors of emission spectra
there is no parallelism between protonation and complexation
with nontransition metal ions.

The complexation of aromatic ligands with transition metal
ions usually produces electronic spectral shifts which are much
greater than the shifts produced by complexation of the same
ligands with nontransition metal ions. For example, the lowest
frequency $\pi \rightarrow \pi^*$ absorption band of the 2:1 complex of 8-hydroxy-
quinoline with Copper (II) lies at 2.44×10^4 cm^{-1}. This is the
result of a secondary type of π-bonding in which charge is
transferred from d orbitals on the transition metal ion to
vacant π-orbitals on the ligand. This intramolecular charge-
transfer interaction stabilizes the excited states relative to
the ground state and accounts for the very low frequency absorption spectra. Most transition metal ion complexes of aromatic ligands do not fluoresce or phosphoresce even though the
ligands do. In this respect, the complexation by the transition
metal ion may be said to result in static quenching of the ligand

luminescence. The reasons for the failure of transition metal
complexes to fluoresce are not completely understood. One theory
maintains that the presence of the paramagnetic transition metal
ion causes quenching by the reversible electron transfer mechanism. Others, maintain that the paramagnetic and heavy atom
effects of the transition metal ion cause spin orbital coupling
which populates low lying high multiplicity states which are
then deactivated by internal conversion.

In a few cases luminescences from transition metal complexes
have been observed. The 3:1 complex of 8-hydroxyquinoline or
acetylacetone with Cr(III) phosphoresces as a result of a ligand
field transition (38). In this case, the light energy is absorbed by the aromatic ligand but the luminescence arises from
a transition between states localized on the transition metal
ion (states arising from the d orbital configurations). Several
complexes of aromatic ligands with lanthanide elements, notably
Eu(III) also absorb through the ligands but the excitation
energy is transferred to the metal ion and fluorescence arises
as a result of transitions between the f orbitals of the lanthanide ion. The bandwidth characteristics of d-d and f-f
luminescences are interesting in that they are very narrow,
almost line-like. This is a consequence of the transitions
arising from states localizing on the metal ions so that they
are not much affected by the vibrational structure of the whole
molecule. Complexes of 2,2'-bipyridyl with ruthenium (11) are
unusual in that they luminesce from a charge-transfer state
(39). The bandwidth of this charge-transfer luminescence is
broader than the d-d and f-f emissions of the ion localized
luminescences, but considerably narrower than typical emissions
arising from π,π^* states.

Concentration Effects

The arguments developed, up to the present, for the spectral behavior of molecules in solution were based upon the
behavior of molecules in very dilute solutions ($< 10^{-4}$ M) where

the interactions were predominantly of the solute-solvent type.
High concentrations of absorbing or luminescing species, in
solution, may cause problems in the interpretations of molecular
electronic spectra. Some of these problems arise from instrumental considerations and must be dealt with from a practical
point of view. However, some aspects of the spectroscopy of
concentrated solutions arise from solute-solute interactions
and their consideration is fundamental to the understanding of
chemical spectroscopy.

It is well known that in the limit of high absorber concentration, the linearity between absorbance and concentration
breaks down. Usually, this is observed as a diminution of the
rate of increase of absorbance with increasing absorber concentration. In very concentrated solutions the absorbance may
become infinite because all of the exciting light will be absorbed before it can completely penetrate the sample cell.
However, at concentrations well below the limit of total absorption, the nonlinearity of absorbance with concentrations
is due to a chemical phenomenon in the ground electronic state.
As the solute concentration is raised, the frequency of en-
counters between the solute molecules is increased. Often the
solute molecules are capable of forming polymolecular aggregates
or complexes in which the aromatic systems are coupled. When
this phenomenon occurs, the complex species are formed at the
expense of the monomeric species whose absorption is being
monitored, usually at a fixed wavelength. If the aromatic
systems of the monomers are coupled in the aggregates, the
complete absorption spectrum will usually reveal a shift to
lower frequency resulting from the absorption by the polymeric
species. However, even if the aromatic systems of the monomers
comprising the aggregates are not coupled appreciably so that
no red shift occurs, the effective concentration of absorbers
is still diminished even though absorption still occurs princi-
pally at the original wavelength of measurement.

The intensities of fluorescence and phosphorescence also

demonstrate deviations from linearity with increasing absorber
concentration. However, in the cases of the luminescence processes, the deviations from linearity usually become obvious
at absorber concentrations much lower than those that will produce nonlinearity in the dependence of absorbance on concentration (i.e. $< 10^{-2}$ M). In most instances, in solutions where
the concentrations of absorber are high enough to cause nonlinearity of absorbance, the intensities of fluorescence and
phosphorescence actually decrease with increasing absorber
concentration (Fig. 2.6).

In the lower end of the absorber concentration region where
fluorescence and phosphorescence intensities show deviations
from linearity (i.e. $\sim 10^{-4}$ M for most $\pi^* \rightarrow \pi$ transitions), the
nonlinearity of the luminescence signal is not due to molecular
interaction of any type but is the consequence of the fact that
the intensities of fluorescence and phosphorescence do not truly
vary linearly, but rather exponentially, with absorber concentration. Only in the limit of extremely low absorber concentration (and low molar absorptivity in excitation) where higher
than first order terms in Eqn. (1.10) are negligible, does the
expression for fluorescence intensity (or the expression for
phosphorescence intensity) vary linearly with absorber concentration. A solution with an absorbance of 0.1 at the wavelength
of excitation, for example, will have a fluorescence intensity
10% lower than that predicted by Eqn. (1.12).

At somewhat higher concentrations, solute-solute interactions may add to the nonlinearity of the luminescence intensity
vs absorber concentration curve. However, the concentration
range in question is still below that in which aggregation
effects are observed in the absorption spectrum and implies that
the solute-solute interactions are occurring in the lowest excited singlet state. Two types of excited state solute-solute
interaction are common. The aggregation of excited solute
molecules with ground state molecules of the same type produces
an excited polymer or "excimer", which by virtue of the coupling
of the aromatic systems of the excited and unexcited molecules

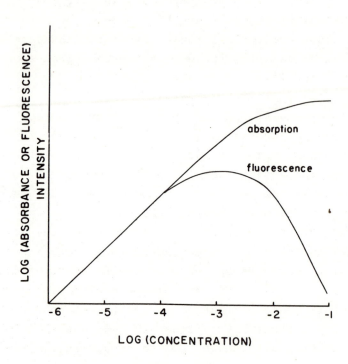

Fig. 2.6. Illustrating the dependence of absorbance and
          fluorescence intensity on analyte concentration.

usually luminesces at lower frequency than the monomeric excited
molecule (although quenching sometimes occurs as a result of
excimer formation). For example, at 10^{-4} M, pyrene exhibits a
structured fluorescence in the near ultraviolet, as the concentration of pyrene is brought to 10^{-2} M the structured ultraviolet
fluorescence diminishes and a new unstructured emission in the
blue region of the spectrum appears (40). The intensity of the
blue fluorescence is proportional to the square of the pyrene
concentration and has thus been attributed to a pyrene dimer
formed in the excited state. Occasionally, excited state complex formation may occur between two different solute molecules.
The phenomenon is comparable to excimer formation, however, the
term "exciplex" has been coined to describe a heteropolymeric
excited state complex. Excimer and exciplex formation are
usually observed only in fluorescence phenomena because diffusion of the excited species is necessary to form the excited
state complexes. However, one concentration effect that is
observed in molecules in fluid or rigid media (i.e. occurs in
both fluorescence and phosphorescence phenomena) is energy
transfer. Energy transfer entails the excitation of a molecule
which during the lifetime of the excited state passes its excitation energy off to a nearby molecule. The loss of excitation energy from the species excited, which may be of analytical
interest, results in quenching of the luminescence of the energy
donor and may result in luminescence from the energy acceptor
species which becomes excited in the process.

Energy transfer from an excited donor molecule to an unexcited acceptor molecule commonly occurs by one of two alternative
concentration dependent processes. In the resonance excitation
transfer mechanism (40,41), also called the dipole or Förster
mechanism, the donor and acceptor molecules are generally not
in contact with one another and may be separated by as much as
$7 \times 10^{-7} - 1 \times 10^{-6}$ cm. In the classical sense, the excited
energy donor molecule may be thought of as an electrical dipole
which creates an electrical field in its vicinity. Potential
acceptor molecules within the range of this electrical field

may absorb energy from the field resulting in electronic excitation of the latter. Because the energy of excitation of the acceptor comes from the excited donor, the donor is returned, radiationlessly, to its ground electronic state. The acceptor molecule which is then excited, may return to its ground state by luminescence (sensitized luminescence) or by any of the various radiationless pathways for the deactivation of electronically excited states. The probability and thus the rate of resonance energy transfer decreases as the sixth power of the distance between the donor and acceptor dipoles according to

$$k_{ET} = \frac{1}{\tau_D} \left(\frac{R_o}{R} \right)^6 \qquad (2.42)$$

where k_{ET} is the rate constant for resonance energy transfer, τ_D is the lifetime of the excited state of the donor molecule, R is the mean distance between the centers of the donor and acceptor dipoles and R_o is a constant for a given donor-acceptor pair (the critical separation) corresponding to the mean distance between the centers of the donor and acceptor dipoles for which energy transfer from donor to acceptor and luminescence from the donor are equally probable. When $R_o > R$, energy transfer is more probable than luminescence from the donor. When $R_o < R$ most of the excited donor molecules will luminesce rather than transfer excitation energy to the acceptor species. Experimentally,

$$R_o = \left(\frac{3000}{4\pi N [A]_{1/2}} \right)^{1/3} \qquad (2.43)$$

which reduces the quantum yield of fluorescence of the donor to half of its value measured in the absence of the acceptor species. R_o and $[A]_{1/2}$ are therefore the two important constants of any donor-acceptor pair between which energy is transferred by the

resonance mechanism. The higher the concentrations of both
donor and acceptor, the smaller is R and the more probable
resonance energy transfer becomes. Resonance energy transfer
is thus seen to be a concentration dependent phenomenon. The
rate of resonance energy transfer is also determined by the
lifetimes and therefore the spin multiplicities of the excited
electronic states, of the donors as well as by the probabilities
(i.e. molar absorptivities) of the electronic transitions involved in the acceptors. Donor molecules in the lowest excited
singlet state have short lifetimes whereas donors in the triplet
state have long lifetimes. On the other hand, transition to
the lowest excited singlet state usually has a high probability
whereas transition from the ground state to the lowest triplet
state has very low probability. The high probability of singlet
singlet transition in most acceptor molecules usually compensates
for the short lifetime of the excited donor, so that singlet-
singlet excitation of acceptors by excited singlet donors (singlet-singlet energy transfer) is a probable phenomenon commonly
observed as singlet sensitized fluorescence. The long lifetime
of the triplet state and the high probability of singlet-singlet
transition in an acceptor also make triplet-singlet resonance
energy transfer, as observed in the sensitization of acceptor
fluorescence by triplet donors, a fairly common phenomenon.
In cases of triplet-singlet energy transfer, the fluorescence
of the acceptor will generally exhibit a lifetime comparable to
that of the triplet donor (i.e. much longer than that observed
when the acceptor is excited directly). However, because of
the very low probability of singlet-triplet transitions in
acceptor molecules, both singlet-triplet and triplet-triplet
resonance energy transfer have extremely low probabilities, as
a result of which, singlet or triplet sensitized phosphorescences originating from the resonance energy transfer mechanism
are not generally observed. Another general requirement for
the occurrence of resonance energy transfer is the overlap of
the fluorescence (or phosphorescence) spectrum of the donor and
the absorption spectrum of the acceptor. Any degree of overlap
of these spectra will satisfy the quantization requirements for

the energy of the thermally equilibrated donor molecule to promote the acceptor to a Franck-Condon excited singlet state.
However, the greater the degree of overlap of the luminescence
spectrum of the donor and the absorption spectrum of the
acceptor, the greater is the probability that energy transfer
will take place.

The exchange mechanism of excitation energy transfer (43)
is a shorter range phenomenon than resonance energy transfer,
the rate of energy transfer decreasing exponentially with decreasing acceptor concentration. This mechanism becomes important only when the electron clouds of donor and acceptor are
in direct contact (i.e. the donor and acceptor molecules are
separated by a distance no greater than their collision diameter).
In this circumstance, the highest energy electrons of the donor
and acceptor may, as a result of their indistinguishability,
change places. Thus the optical electron of, say, a triplet
donor molecule may become part of the electronic structure of
an acceptor molecule originally in the ground singlet state
while the donor is returned to its ground singlet state by
acquiring an electron from the acceptor. As in the case of
resonance energy transfer, exchange energy transfer is most
efficient when the equilibrated excited state of the donor lies
slightly higher than the Franck-Condon excited state of the
acceptor. However, whereas the resonance mechanism precludes
energy transfer resulting in a spin forbidden transition of
the acceptor molecule (i.e. singlet-triplet and triplet-triplet
energy transfer do not occur appreciably by the resonance
mechanism), triplet-triplet energy transfer commonly, and
singlet-triplet energy transfer occasionally occur by the exchange mechanism, provided that the donor triplet in the former
case and the donor singlet in the latter case are higher in
energy than the acceptor triplet state. Singlet-singlet and
triplet-singlet transfer are also permitted by the exchange
mechanism. Exchange energy transfer is a diffusion limited
process (i.e. every collision between donor and acceptor leads
to energy transfer) and as such its rate depends upon the

viscosity of the medium. Resonance energy transfer, on the other
hand, is not diffusion limited, does not depend upon solvent
viscosity and may be observed at lower concentrations of acceptor
species. In fluid solutions of relatively low viscosity (e.g.
water, ethanol, cyclohexane) diffusion limited rate constants
are of the order of $\sim 10^{10}$ mole^{-1} sec^{-1}. The lifetime of the
lowest excited singlet state of a donor molecule is $< 10^{-7}$ sec.
Thus if $[A]_{1/2}$ for a given acceptor species is less than 10^3 M,
singlet-singlet and singlet-triplet exchange energy transfer is
too slow to compete with fluorescence for deactivation of the
excited donor. On the other hand, the lifetime of the lowest
triplet state of most triplet donors, in fluid solution, is
$> 10^{-4}$ sec. As a result, if $[A]_{1/2}$ for a given acceptor species
is greater than 10^{-4} M, virtually all triplet donor species will
be deactivated. This accounts for the high efficiency of oxygen
quenching of triplet species (e.g. biacetyl) in fluid solutions.
In rigid solutions, where phosphorescence is usually observed,
exchange energy transfer is normally precluded unless the donor
and acceptor species form a contact complex. However, in some
solid systems, especially in organic crystals containing trace
impurities (acceptors), where there is a high degree of order
in the matrix, energy transfer may be observed as a result of a
process known as exciton migration. In this process the energy
absorbed by the donor is delocalized throughout the matrix and
is eventually localized by the acceptor species which then emits
fluorescence or phosphorescence. This process is observed in
anthracene crystals doped with a small amount of tetracene.
Excitation of the crystal at frequencies where only anthracene
absorbs results in fluorescence from tetracene (44).

Because the efficiency of resonance energy transfer depends
upon the distance between relatively widely separated donor and
acceptor moieties, this process has found utility in the measurement of the distances between fluorophores in protein molecules
(45). Exchange energy transfer, has, to date, found its
greatest utility in the efficient sensitization of multiplicity
forbidden luminescence in certain rare earth chelates (46-48).

This phenomenon is produced by exciting the ligand of the chelate
to its lowest excited singlet state. The ligand then transfers
its excitation energy to the lanthanide ion, causing promotion
of an f electron to a higher f orbital. Luminescence then ensues
from the lanthanide ion. This process has resulted in the development of several rare earth based phosphors and lasers. From
the analytical point of view, energy transfer may be regarded
primarily as an interference although a limited amount of analytical methodology has been derived from the phenomenon.

At very high absorber concentrations, ground state aggregation effects may come into play and may ultimately result in
the decrease of luminescence intensity with increasing absorber
concentration. Moreover, when very highly absorbing solutions
are studied, the loss in luminescence intensity with increasing
absorber concentration may also be due to absorption of all of
the exciting light before it completely traverses the sample.
Luminescence is then observed from only part of the sample cell.
If the absorption spectrum and the fluorescence spectrum of the
solute overlap, diminution of fluorescence intensity may result
from the reabsorption of part of the emitted radiation. This
effect is called trivial reabsorption and is observed as a
concentration dependent effect in fluorescence spectroscopy but
not in phosphorescence spectroscopy because of the lack of
overlap of the absorption and phosphorescence spectra.

Fluorescence and Phosphorescence of Proteins and Nucleic Acids

The proteins and nucleic acids are of great interest to
biological scientists because of their pervasiveness in living
systems. These macromolecules are comprised of copolymerized
amino acids (the proteins) or nucleotides (the nucleic acids)
some of which, in the monomeric form, absorb near ultraviolet
light and fluoresce or phosphoresce. Inclusion of the monomeric
chromophores into the macromolecular structure of a protein or
nucleic acid results in special aggregation effects with the
result that the absorption or luminescence properties of the
macromolecules are not simply the sum of the optical properties

of the component monomers.

The proteins contain three amino acids, phenylalanine, tyrosine and tryptophan which have their long wavelength absorption maxima at 250-300 nm. These amino acids are also fluorescent, phenylalanine fluorescing maximally at 282 nm (ϕ_f = 0.04), tyrosine at 303 nm (ϕ_f = 0.21) and tryptophan at 348 nm (ϕ_f = 0.19), and are responsible for the native fluorescence of proteins (49). Because of the small molar absorptivity and quantum yield of phenylalanine, the fluorescence of phenylalanine, in proteins, is negligible by comparison with those of tyrosine and tryptophan in the same protein. Even in proteins containing large phenylalanine to tyrosine or phenylalanine to tryptophan ratios phenylalanine emission is negligible, an observation which suggests that energy transfer from phenylalanine to tyrosine or tryptophan, within the protein molecule, may be very efficient. Proteins which contain tyrosine but no tryptophan, generally exhibit the fluorescence of tyrosine alone (50). However, if tryptophan is present in the protein, even in small amounts, its fluorescence dominates that of the protein. In most (though not all (51)) cases, protein fluorescence is exclusively that of tryptophan (50). This phenomenon does not occur in dilute equimolar solutions of the three aromatic amino acids and suggests that the structure of the protein is responsible for the quenching of phenylalanine and tyrosine fluorescence. Resonance energy transfer between constituent aromatic amino acids probably accounts for most of the observed protein fluorescence phenomena. The proximity between aromatic amino acids arising from the sequence of amino acids in the peptide chain and more important, from the folding of the peptide chain as a result of hydrogen bonding, assures that in many cases the distance between aromatic amino acid residues will be within the 7×10^{-7} - 1×10^{-6} cm limit of appreciable dipole-dipole interaction. This is supported by the loss in intensity of tryptophan fluorescence and gain in intensity of tyrosine fluorescence which occurs when proteins are denatured (i.e. the intramolecular hydrogen bonds are broken), by urea, acid or alkaline solutions.

The nucleic acids DNA and RNA consist of copolymerized
nucleotides, each nucleotide consisting of an amino or hydroxyl
substituted purine or pyrimidine base, condensed with a five-
membered sugar (ribose or 2'-deoxyribose), which is in turn
esterified with phosphoric acid. Nucleotides differ according
to their purine or pyrimidine bases. The most important nucleotides in biological polymers are adenylic acid, guanylic acid,
cytidylic acid and uridylic acid, which are the component nucleotides of RNA having as purine and pyrimidine bases adenine,
guanine, cytosine and uracil, as well as deoxyadenylic acid,
deoxyguanylic acid, deoxycytidylic acid and deoxythymidylic
acid, the component nucleotides of DNA having adenine, guanine,
cytosine and thymine as bases. The polynucleotides or nucleic
acids form by pyrophosphate condensation (dehydration polymerization). The ribonucleic acid and some deoxyribonucleic
acid polymers are single stranded. However, the most important
form of DNA consists of two intertwined α-helicies, held together
by interhelical hydrogen bonds between acidic hydrogen atoms and
basic sites on purine-pyrimidine base pairs. The optical properties of nucleic acids in the native state arise from the
purines and pyrimidines which have long wavelength absorption
maxima between 250 and 275 nm and luminesce from several ionic
forms. A rather interesting phenomenon associated with the
absorption spectra of the bases in RNA and DNA is hypochromicity.
It might be expected that the absorbance of a DNA solution at
some wavelength in the region 250-275 nm (where all of the bases
absorb) would be equal to the absorbance, at the same wavelength,
of a solution containing equimolar amounts of the nucleotides
comprising the polynucleotide. Actually, the absorbance of the
polymer is generally found to be smaller (or hypochromic) than
the absorbance of the solution of randomly oriented monomers.
The hypochromism of nucleic acids is attributed to the proximity
of identical absorbing groups in the polymer. The effect of
proximity is the vectorial coupling of the transition moments of
each chromophore, resulting in a net decrease of the average

transition moment per chromophore when averaged over all chromophores of a given type in the polymer, relative to the average
transition moment per monomer. Because the intensity of absorption (the absorbance) is proportional to the square of the
transition moment, the absorbance of the polymer is smaller than
that of a solution containing equivalent amounts of each absorbing monomer. Hypochromism due to aggregation may also contribute
to the deviation from Beer's law, of simple absorbing species,
in solution, at high temperatures.

   The fluorescences and phosphorescences of the nucleic acids
are not well understood, their study being in a relatively
primitive state. This state of affairs is due to the fact that
the emissions of the purines, pyrimidines, nucleotides and nucleic acids themselves, are weak, especially at ambient temperature.
Also, because of the large number of functional groups on each
purine or pyrimidine, absorption and emission from several tau-
tomers derived from each species is possible. As a result, there
is often disagreement over whether emissions reported in the
literature are authentic or derived from impurities. It is now
generally agreed that DNA fluoresces with maximum near 355 nm.
The fluorescence of DNA differs considerably from that of any
of its component bases or nucleotides, lying at $\sim$ 5000 cm^{-1} to
lower frequencies. The fluorescence of DNA, however, does bear
a strong resemblance to the fluorescences of the dinucleotides
and polynucleotides, diadenylic acid and polyadenylic acid (λ_f
= 350 nm) and adenylthymidylic acid and polyadenylthymidylic
acid (λ_f = 355 nm). These di- and polynucleotides are believed
to form exciplexes (52) between base pairs, in their excited
states, thereby accounting for the long wavelengths of fluorescence relative to those of the monomers. Apparently, exciplexes
between adjacent bases in DNA account for the long wavelength of
fluorescence of the biological macromolecule. It has been shown
that interhelical hydrogen bonding between guanine and cytosine
moieties quenches the fluorescence of exciplexes derived from
adenine and thymine contribute to the fluorescence of DNA.
Although energy transfer might be expected to dominate the

luminescence of DNA, in which case the fluorescence of guanine, having the lowest first excited singlet state of the bases, should represent the fluorescence of DNA, this is apparently not the case. Energy transfer in native DNA if it occurs at all is not an important process in the lowest excited singlet state. On the other hand, the phosphorescences of polynucleotides are similar to the phosphorescences of the free bases and show no evidence for triplet exciplex formation. The unstructured phosphorescence of DNA at ∿ 450 nm has been attributed to the lowest triplet state of thymine which is presumably populated, in large part, by energy transfer from the other bases. However, many of the details of the phosphorescence of DNA are poorly understood. For example, it is not known whether intersystem crossing occurs in the bases which are directly excited, followed by exchange energy transfer to and phosphorescence from thymine, or whether resonance or exchange singlet-singlet transfer from the bases directly excited to the thymine moiety, followed by intersystem crossing, is responsible for populating the lowest triplet state of the thymine residue.

The binding of small fluorescent molecules either covalently or by electrostatic association is a subject which has received considerable attention in recent years (53), and is of considerable importance in molecular pharmacology, biopharmaceutics and biochemistry. Binding may either quench or enhance the fluorescent emission of the bound fluorescent probe. In cases where enhancement or incomplete quenching of the fluorescence of the substrate occurs upon binding, the position of the fluorescence maximum of the bound molecule is often shifted relative to the position of the fluorescence maximum of the unbound species.

The binding of 8-anilino-1-naphthalenesulfonate to bovine serum albumin has been studied extensively. The increase in the quantum yield of fluorescence and the shifting of the fluorescence maximum to shorter wavelengths, of 8-anilino-1-naphthalenesulfonate, upon binding to albumin, have been analogized to the corresponding spectral changes which occur when this compound

is transferred from aqueous to low dielectric media. This, it
has been suggested, indicates that the binding entails hydro-
phobic interior of the protein molecule. The covalent probe
5-dimethylamino-1-naphthalenesulfonyl chloride which is used
to label N-terminal amino groups in proteins shows similar
changes in its fluorescence upon binding.

Whereas proteins, at biological pH, tend to form strong
complexes with anionic, cationic and neutral ligands, regardless
of the charge on the proteins, the nucleic acids form strong
complexes with only cationic ligands, at biological pH, because
of their negative charge, due to the presence of the anionic
phosphate ester linkages. The singly charged cation derived
from acridine orange (3,6-bis-dimethylaminoacridine) binds to
DNA, causing changes in the fluorescence spectrum which are
similar to those observed upon going from dilute ($\sim 10^{-6}$ M) to
concentrated (> 10^{-4} M) aqueous solutions of the ligand (54).
The changes in the acridine orange monocation fluorescence upon
binding to DNA have been interpreted to indicate the occurrence
of stacking interactions (aggregation) between the individually
bound ligands, some of which are sandwiched (intercalated) between base pairs of the DNA helix and others of which are bound
to the sugar-phosphate "backbone" of the biopolymer.

The binding of small fluorescent molecules to proteins and
nucleic acids is a field of great potential importance in biological science. However, the state of the art is such that
only a small number of fluorescent probes have been investigated.
Hence, it would probably be premature to advance generalizations
about the nature of binding interactions or about their influences upon the luminescence spectra of the bound ligands.

Polarization of Fluorescence and Phosphorescence

Light consists of traveling waves each of which has an
associated electric field whose vector oscillates in a plane
perpendicular to the direction of travel of the light wave.

There is also a magnetic field associated with the light wave, whose vector oscillates in a plane perpendicular to both the direction of travel and the plane of oscillation of the electric field vector. It is predominantly the interaction of the electric field vector of exciting light with the π-electrons of the absorbing molecule which changes the electronic dipole moment of the mclecule and thereby produces the electronic configuration of the excited molecule. Because the absorptive transition consists of the interaction of two vectors, the orientation of the electric vector of the exciting light relative to the direction of the transition moment vector (the line along which charge is displaced in the molecule during transition) is of prime importance in determining the magnitude of interaction. The square of the scalar projection of the transition moment upon the electric field vector is proportional to the intensity of light absorbed (I_{abs}) in the transition:

$$I_{abs} \propto (M\cos\theta)^2 \qquad (2.44)$$

where θ is the angle between the transition moment vector $\vec{M}$ of the molecule, and the electric vector of the light (Fig. 2.7).

In ordinary unpolarized light, there is a spherical distribution of electric vectors, of the different component waves, about the direction of travel of the waves. This is a result of the random orientations of the individual emitters in the light source. Consequently, it is difficult to speak of directional relationships between the molecular transition moments and the electric vectors of the unpolarized exciting light waves. If, however, ordinary light is passed through a polarizing device such as an oriented polymer film, all light waves are theoretically filtered out except those that have their electric vectors oscillating in one plane. In this case, the intensity of the incident polarized light, of a given frequency, absorbed by each molecule experiencing electronic transition, is jointly proportional to the square of the magnitude of the transition moment $\vec{M}$, belonging to the transition excited, and to the square

of the cosine of the angle Θ between the direction of the transition moment vector and the plane of polarization of the exciting
light (i.e. the direction of the electric vector of the exciting
light), according to Eqn. (2.44). In other words, those molecules
whose transition moments make the smallest angles with the electric field vector of the exciting light absorb the greatest
amount of light.

Now let us consider the emission of fluorescence or phosphorescence from a molecule which has absorbed polarized light.
The intensity of light emitted in the direction of the luminescence transition moment, I_f, is proportional to the square of
the magnitude of the transition moment according to

$$I_f \propto M^2 \qquad (2.45)$$

where M is the magnitude of a vector which may be resolved into
components projected in the direction of observation, parallel
to the electric vector of the exciting light, and in the direction of travel of the exciting light (the y, z and x axes,
respectively, in Fig. 2.7). The latter two components are of
interest to us here.

The component of $\vec{M}$ which is parallel to the electric vector
of the exciting light will hereafter be referred to as $\vec{M}$,
while that pointed in the direction of travel of the exciting
light (perpendicular to the electric vector of the exciting
light) will be referred to as $\vec{M}$. If Θ is the angle between $\vec{M}$
and the direction of the electric vector of the exciting light,
and if ϕ is the azimuthal angle $\vec{M}$ makes the direction of observation, then

$$M_{\parallel} = M\cos\Theta \qquad (2.46)$$

and

$$M_{\perp} = M\sin\Theta\sin\phi \qquad (2.47)$$

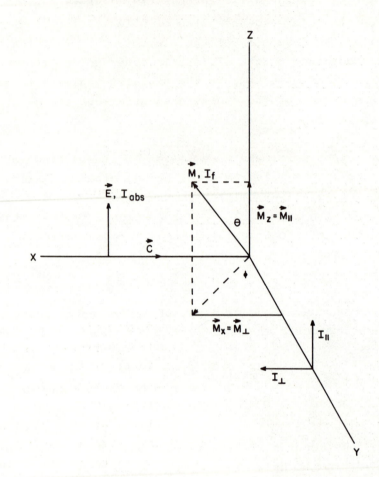

Fig. 2.7. Illustrating the geometrical arrangement of a photo-
          selection (polarized emission) measurement. Exciting
          light with intensity I_{abs}, and electric vector $\vec{E}$,
          polarized in the XZ plane and traveling in the positive
          X direction (with velocity $\vec{C}$) impinges on an array of
          potentially absorbing and fluorescing molecules, pro-
          ducing electronic excitation. Any given molecule has a
          transition moment $\vec{M}$ for fluorescence, which makes an
          angle Θ with the direction of polarization of $\vec{E}$ (the
          Z direction). Fluorescent light is emitted along the
          line of $\vec{M}$, and is resolved into components parallel
          ($I_{\parallel}$) or perpendicular ($I_{\perp}$) to the direction of polar-
          izing of $\vec{E}$ by means of polarizing and analyzing ele-
          ments having their optical axes parallel and perpendi-
          cular to each other, respectively.

The light intensities emitted with electric vectors parallel to and perpendicular to the electric vector of the exciting light are $I_\parallel$ and $I_\perp$, respectively, and are given by

$$I_\parallel \propto M_\parallel{}^2 = M^2\cos^2\theta \qquad (2.48)$$

and

$$I_\perp \propto M_\perp{}^2 = M^2\sin^2\theta\sin^2\phi \qquad (2.49)$$

The polarization (or degree of polarization) p of the emission is defined by

$$p = \frac{I_\parallel - I_\perp}{I_\parallel + I_\perp} \qquad (2.50)$$

The luminescence signals measured with the optical axis of the polarizer in front of the exciting source, aligned parallel to that of the polarizer in front of the detector, and then with the optical axes of the two polarizers aligned perpendicular to each other, are proportional to $I_\parallel$ and $I_\perp$, respectively, and p can thus be calculated at any excitation and emission wavelength combination from simple measurements. The significance of p is that it is a measure of the extent to which the emitted light has its electric vector pointing in the same direction as that of the exciting light (i.e. it is the degree of retention of polarization by the emitted light). If the polarization, p, of the emission maximum is plotted as a function of the wavelength or frequency of the exciting light, the polarization of the fluorescence excitation spectrum or the polarization of the phosphorescence excitation spectrum is obtained.

Theoretically, p can take on values from +1 to -1. Values of p approaching these theoretical extremes are sometimes observed in the emission polarization of oriented polymers (i.e.

samples with the molecules all lined up in the same direction).
In these systems, the absorption moments of all absorbing chromophores can be almost perfectly aligned with the direction of the
electric vector of the exciting light. For absorptions with
moments coincident with the emission moment, $I_f = I_{\parallel}$, and so
p = +1. For angles between the absorption and emission moments,
between 0° and 90°, p is intermediate between +1 and -1.

In actual practice, in solution, the molecules are randomly
oriented so that the values of $I_{\parallel}$ and $I_{\perp}$ must be averaged over
all values of Θ and ϕ between 0° and 90°. This treatment, with
the approximation that the molecules in solution are rigid and
do not interact, yields maximal values of p (designated p_o) of
+1/2 for coincident absorption and emission transition moments
of the molecule and -1/3 for perpendicular absorption and emission transition moments, where p_o denotes the polarization in
the absence of molecular motion or interaction with other molecules. If the absorption and emission transition moments are
suituated at an angle β, with respect to one another, the
polarization p_o is denoted by (55)

$$p_o = (3\cos^2\beta - 1)/(\cos^2\beta + 3) \qquad (2.51)$$

In the fluorescence excitation polarization spectrum, the
most positive values of p_o will be observed at or near the
excitation maxima of the transitions having their absorption
moments pointing in the same (or nearly the same) direction as
the fluorescence transition moment. The most negative values of
p_o will occur at or near the excitation maxima corresponding to
transitions having their absorption transition moments perpendicular (or nearly so) to the fluorescence transition moment.
It is to be expected that direct absorption to the state from
which fluorescence occurs would show the highest positive
polarization because the absorption and fluorescence would then
correspond to the same transition in opposite directions. This
aspect of fluorescence excitation polarization is invaluable in
the assignment of emission bands to the states from which they

arise, and in the location of weak absorption bands buried under
more intense absorption bands. For example, polarization measurements revealed the existence of a hidden high energy n → π*
absorption buried under the tailing of the lowest π → π* absorption band of sym-tetrazine (56). Similar arguments are also
applicable to the polarization of phosphorescence. Here, however,
the situation is more complicated because the states to which
absorption occurs are never the states from which emission occurs,
and a substantial component of the phosphorescence transition
moment lies perpendicular to the molecular plane, whereas in
fluorescence and absorption the transition moments lie in the
molecular plane. The detailed consideration of the polarization
of phosphorescence is beyond the scope of this book.

Up to the present, the absorbing and emitting molecules
have been considered to maintain fixed positions in solution,
and not to interact with one another. These approximations are
valid in glassy solutions or very viscous solvents where molecular rotations are non-existent or very slow compared with the
lifetime of the luminescent molecule and in very dilute solutions where solute-solute interactions are improbable. In such
solutions, polarization values measured will approach the
theoretical maxima of p_o. In solvents of low viscosity or of
high solute concentrations, the values of p measured are always
found to be lower than p_o.

In the case of low solvent viscosity, the diminution of p
is the result of appreciable rotation of the emitting molecule
during the lifetime of the excited state, with accompanying loss
of molecular orientation and hence loss of polarization of the
emission. In the case of high solute concentration, the diminution of p is due to transfer of the energy of excitation to
another molecule with a different orientation. Emission then
occurs from the second molecule. Again the result is an apparent
loss of polarization of the emitted radiation.

The rotational depolarization may be studied apart from the

depolarization due to energy transfer by observation of the
emission of solutes in solvents of varying viscosity and of low
solute concentration. Similarly, the depolarization due to
energy transfer may be studied apart from the rotational effect
by observation of the emission of solutions of varying concentrations of solute in viscous or rigid media. Because phosphorescence occurs almost exclusively in rigid media, phosphorescence depolarization studies are limited to energy transfer
phenomena.

In the absence of energy transfer, the polarization is
related to rotational diffusion phenomena according to (57)

$$p = [(\frac{1}{p_o} - \frac{1}{3}) (1 + \frac{RT}{\eta} (\frac{\tau}{V})) + \frac{1}{3}]^{-1} \qquad (2.52)$$

where R, T and η are, respectively, the gas constant, the absolute temperature and the viscosity of the medium, and τ and V
are, respectively, the lifetime of the excited state and the
molecular volume of the solute. Consequently, if R, τ and η
are known and p_o is determined by measurement in rigid media,
measurement of p yields values of τ/V. If the molecular volume,
in solution, can be estimated, Eqn. (2.52) can be used to calculate approximate values of τ. On the other hand, if τ can be
measured independently and accurately, as is currently possible
with modern instrumentation, values of the molecular volume in
solution may be determined which are accurate within the limits
of the assumptions upon which Eqn. (2.52) is based. Eqn. (2.52)
is based upon a spherical molecular model. A more elaborate,
ellipsoidal model is also available (58,59).

Depolarization due to energy transfer occurring from non-
collisional processes (i.e. Förster energy transfer) and in the
absence of rotational depolarization may be calculated from (60)

$$p = \left[\left(\frac{1}{p_o} - \frac{1}{3} \right) \left(1 + \frac{105NR_o^6C}{a_m^3} \right) + \frac{1}{3} \right]^{-1} \qquad (2.53)$$

where N is Avogadro's number, a_m the effective molecular radius, R_o the distance between dipoles at which the probability of emission is equal to the probability of energy transfer and C is the concentration in moles/liter. Studies of polarization dependence on energy transfer can thus be useful in determining molecular dimensions or quantities of spectroscopic interest, such as R_o.

REFERENCES

1. Bayliss, N. S. and McRae, E. G., J. Phys. Chem. 58, 1002 (1954).

2. Schulman, S. G. and Capomacchia, A. C., Anal. Chim. Acta 58, 91 (1972).

3. Krishna, V. G. and Goodman, L., J. Amer. Chem. Soc. 83, 2042 (1961).

4. Schulman, S. G. and Sanders, L. B., Anal. Chim. Acta 56, 91 (1971).

5. Werner, T. C. and Hercules, D. M., J. Phys. Chem. 73, 2005 (1969).

6. Jackson, G. and Porter, G., Proc. Roy. Soc. A260, 13 (1961).

7. Mataga, N., Kaifu, Y. and Koizumi, M., Bull. Chem. Soc. Japan 29, 373 (1956).

8. Brederek, K., Forster, T. and Oesterlin, H. G., Luminescence of Organic and Inorganic Materials, H. P. Kallmann and G. M. Spruch, Eds., Wiley, New York, 1962, p. 161.

9. Goldman, M. and Wehry, E. L., Anal. Chem. 42, 1178 (1970).

10. Schulman, S. G. and Pace, I., J. Phys. Chem. 76, 1996 (1972).

11. Schulman, S. G., Tidwell, P. T., Cetorelli, J. J. and Winefordner, J. D., J. Amer. Chem. Soc. 93, 3179 (1971).

12. Förster, T., Naturwiss 36, 186 (1949).

13. Schulman, S. G. and Winefordner, J. D., Talanta 17, 607 (1970).

14. Schulman, S. G. and Fernando, Q., J. Phys. Chem. 71, 2668 (1967).

15. Weller, A., Z. Elektrochem. 56, 662 (1952).

16. Weller, A., Prog. in React. Kinetics 1, 187 (1961).

17. Schulman, S. G. and Capomacchia, A. C., J. Phys. Chem. 79, 1337 (1975).

18. Schulman, S. G. and Capomacchia, A. C., Spectrochim. Acta 28A, 1 (1972).

19. Schulman, S. G., Capomacchia, A. C. and Tussey, B.,
     Photochem. Photobiol. 14, 733 (1971).

20. Mataga, N., Bull. Chem. Soc. Japan 36, 654 (1963).

21. Levshin, W. L., Z. Physik. 43, 230 (1931).

22. Jaffe, H. H. and Jones, H. L., J. Org. Chem. 30, 964 (1965).

23. Schulman, S. G. and Fernando, Q., Tetrahedron 24, 1777
     (1968).

24. Mason, S. F., Philp, J. and Smith, B. E., J. Chem. Soc. A,
     3051 (1968).

25. Ballard, R. E. and Edwards, J. W., Spectrochim. Acta 20,
     1275 (1964).

26. Capomacchia, A. C. and Schulman, S. G., Anal. Chim. Acta
     59, 471 (1972).

27. Wehry, E. L., J. Amer. Chem. Soc. 89, 41 (1967).

28. Schulman, S. G., Anal. Chem. 44, 400 (1972).

29. Perkampus, H. H., Knop, J. V., Knop, A. and Kossebeer, G.,
     Z. Naturforschung 22a, 1419 (1963).

30. Brinen, J. S., Rosebrook, D. D. and Hirt, R. C., J. Amer.
     Chem. Soc. 87, 2651 (1963).

31. Weller, A., Z. Elektrochem. 60, 1144 (1956).

32. Kovi, P. J., Miller, C. L. and Schulman, S. G., Anal. Chim.
     Acta 61, 7(1972).

33. Yakatan, G. J., Juneau, R. J. and Schulman, S. G., Anal.
     Chem. 44, 1044 (1972).

34. Zalis, B., Capomacchia, A. C., Jackman, D. and Schulman,
     S. G., Talanta, 20, 33 (1973).

35. Aaron, J. J., Winefordner, J. D., Schulman, S. G. and
     Gershon, H., Photochem. Photobiol. 16, 89 (1972).

36. Boaz, H. and Rollefson, G., J. Amer. Chem. Soc. 72, 3435
     (1950).

37. Bhatnagar, D. C. and Forster, L.S., Spectrochim. Acta 21,
     1803 (1965).

38. DeArmond, K. and Forster, L. S., Spectrochim. Acta 19, 1403
     (1963).

39. Lytle, F. E. and Hercules, D. M., J. Amer. Chem. Soc. 91,
     253 (1969).

40. Förster, T. and Kasper, K., Z. Elektrochem. 59, 976 (1955).

41. Förster, T., Ann. Phys. 2, 55 (1948).

42. Förster, T., Z. Naturforsch A4, 321 (1949).

43. Dexter, D. L., J. Chem. Phys. 21, 836 (1953).

44. Bowen, E. J., Mikiewicz, E. and Smith, F., Proc. Phys. Soc. London A62, 26 (1949).

45. Stryer, L. and Haugland, R. P., Proc. Nat. Acad. Sci. U.S. 58, 719 (1967).

46. Weissman, S. I., J. Chem. Phys. 10, 214 (1942).

47. Yuster, P. and Weissman, S. I., J. Chem. Phys. 17, 1182 (1949).

48. Weissman, S. I., J. Chem. Phys. 18, 1258 (1950).

49. Teale, F. W. J. and Weber, G., Biochem. J. 65, 476 (1957).

50. Teale, F. W. J., Biochem. J. 76, 381 (1960).

51. Weber, G., Nature 190, 27 (1961).

52. Eisenger, J., Gueron, M., Schulman, R. G. and Yamane, T., Proc. Nat. Acad. Sci. 55, 1015 (1966).

53. Dandeliker, W. B. and Portmann, A. J., in Excited States of Proteins and Nucleic Acids, R. F. Steiner and I. Weinryb, Eds., Plenum, New York, 1971, p. 199-275, and references contained therein.

54. Van Duuren, B. L., Chem. Revs. 63, 325 (1963).

55. Weber, G., in Fluorescence and Phosphorescence Analysis, D. M. Hercules, Ed., Wiley-Interscience, New York, 1966.

56. Mason, S. F., J. Chem. Soc. 1204 (1959).

57. Perrin, F., J. Phys. Rad. 7, 390 (1926).

58. Weber, G., Biochem. J. 51, 145 (1952).

59. Weber, G., Adv. Protein Chem. 8, 415 (1953).

60. Weber, G., Trans. Faraday Soc. 50, 552 (1954).

BIBLIOGRAPHY

Berlman, I. B., _Energy Transfer Parameters of Aromatic Compounds_, Academic Press, New York, 1973.

Bowen, E. J., Ed., _Luminescence in Chemistry_, Van Nostrand Ltd., London, 1968.

Guilbault, G. G., Ed., _Fluorescence - Theory, Instrumentation and Practice_, Marcel Dekker, New York, 1967.

Guilbault, G. G., Ed., _Practical Fluorescence - Theory Methods and Techniques_, Marcel Dekker, New York, 1973.

Hercules, D. M., Ed., _Fluorescence and Phosphorescence Analysis_, Wiley-Interscience, New York, 1965.

Konev, S. V., _Fluorescence and Phosphorescence of Proteins and Nucleic Acids_, Plenum, New York, 1971.

Mataga, N. and Kubota, T., _Molecular Interactions and Electronic Spectra_, Marcel Dekker, New York, 1970.

Parker, C. A., _Photoluminescence of Solutions_, Elsevier, Amsterdam, 1968.

Pesce, A. J., Rosen, C. G. and Pasby, T. L., _Fluorescence Spectroscopy_, Marcel Dekker, New York, 1971.

Steiner, R. F. and Weinryb, I., Eds., _Excited States of Proteins and Nucleic Acids_, Plenum, New York, 1971.

West, W., Ed., _Chemical Applications of Spectroscopy_, Vol. IX, Part I, Wiley-Interscience, New York, 1968.

Winefordner, J. D., Schulman, S. G. and O'Haver, T. C., _Luminescence Spectrometry in Analytical Chemistry_, Wiley-Interscience, New York, 1972.

Zander, M., _Phosphorimetry_, Academic Press, New York, 1968.

CHAPTER 3
INSTRUMENTATION

Instrumentation for the measurement of fluorescence (Figure 3.1 consists essentially of 1) a light source to electronically excite the sample, 2) a monochromator to separate the light of desired energy from the source, 3) a sample compartment, 4) a second monochromator to isolate the sample's fluorescence energy from the excitation energy, 5) a photodetector to translate the fluorescent light into an electrical signal, and 6) a readout system such as a galvanometer or a recorder, coupled with an amplifier to determine the intensity of fluorescent light emitted. In addition, slits are usually present on either side of the monochromator to collimate the exciting and fluorescent light and to limit the range of wavelengths (bandpass) of exciting light irradiating the sample and fluorescent light falling on the photodetector. Generally, the fluorescence emission detector is placed perpendicular to the exciting beam of light. This configuration allows effective separation of exciting light from the fluorescence emission falling on the photodetector, and in this regard it compliments the monochromators. There are two major types of instruments for fluorimetric measurements; filter fluorimeters and fluorescence (scanning) spectrophotometers. The former uses filters as monochromators while the latter uses gratings to disperse the exciting light and fluorescent emission into their component wavelengths. This will be discussed in detail under wavelength selectors.

Excitation Sources

Since the total fluorescence observed is proportional to the intensity of the source of excitation, an ideal source of exciting light should have the following criteria: i) it should be fairly intense, ii) its output should contain all the wavelengths in the near ultraviolet-visible range, iii) its intensity should be independent of wavelength, i.e. the source should

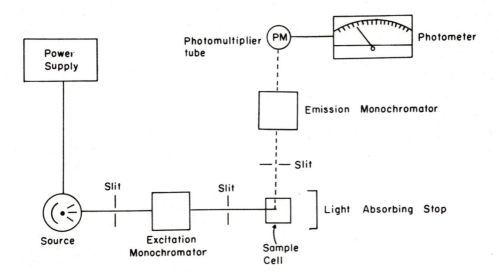

Fig. 3.1.   Schematic diagram of a spectrofluorimeter.

have a smooth continuum radiance, and iv) it should be stable,
i.e. its intensity or spectral characteristics should not change
with time.   There are several types of excitation sources in use.
Selection of a proper source depends upon the nature of the
fluorescence measurement.   Some of the common excitation sources
are discussed below.

Incandescent Lamp.   The most common incandescent lamp
used in spectroscopic studies is the tungsten lamp.   Tungsten
lamps are inexpensive, stable and do not require elaborate power
supplies.   The disadvantages of incandescent lamps are their
limited useful output, covering only the visible range, with
higher radiance in the red region than in the blue, and no output below 300 nm.   Tungsten lamps therefore are useful for excitation in the visible range.   However, since many compounds
are excited by ultraviolet and short wavelength visible radiation, the tungsten lamp has found very little use in fluorescence instrumentation.

Gas Discharge Lamps.   The radiation from gas discharge
lamps originates with the excitation of the gas molecules to
higher electronic states by high-energy intermolecular collisions.   This is achieved by applying a high voltage across an
enclosed gas.   Upon returning to the ground state, ultraviolet
and visible radiation is emitted by the excited molecule.
Common examples of gas discharge lamps are the fluorescent
lights used for indoor illumination and neon signs used in ad-
vertising.   A gas discharge lamp gives out high-energy visible
and near ultraviolet emissions, the wavelength distribution of
which is determined by the nature of the gas filling the lamp.
Each gas emits characteristic wavelengths, usually forming line
spectra, the different line emissions corresponding to different
higher excited states of the gas.   At very low pressure, emissions from any gas are monochromatic.   However, as gas pressure
in the lamp is increased concurrently with the current flowing
through the lamp, the line emissions from the gas are broadened
to bands.   This is known as collisional broadening because the

spectral line broadening results from a smearing of the energy
distribution of the excited gas atoms due to frequent collisions
between the latter. At very high pressure (several atmospheres)
the emissions of a gas discharge lamp broadens out so much as to
make it appear a continuous spectrum. The common gas discharge
lamps used in luminescence spectroscopy are the mercury-vapor
and xenon-arc lamps.

   Mercury-Vapor Lamps. The low pressure mercury-vapor
lamp has intense mercury line emissions, is quite stable and
does not require an elaborate power supply. Because of this,
the mercury-vapor lamp is widely used. Although the mercury-
vapor lamp provides intense emissions at 253·7, 296·5, 302·2,
312·2, 313·2, 365·0, 366·3, 404·7, 435·8, 546·1, 577·0, and
579·0 nm, the most commonly used emissions for spectroscopic
purposes are the group of three lines at 365·0, 365·5, and
366·3 nm. The intensity of the most intense line at 253·7 nm
of a low-pressure mercury-vapor lamp with a quartz jacket is
about 100 times greater than the triplet at about 366.0 nm.
Because of optical consideration which will become apparent
later, the 366 nm triplet is employed in most routine fluorescence spectroscopy, in which the mercury lamp is used. A low-
pressure mercury-vapor lamp, due to its line spectrum, finds
extensive use in the spectrometers utilizing filters for wavelength selection. It is also an excellent light source for
calibration of monochromators. However, because of its line
emission, the mercury-vapor lamp is not useful as an excitation
source for the scanning type spectrofluorimeter or spectro-
phosphorimeter.

   A low-pressure mercury-vapor lamp can be modified to provide energy at wavelengths other than the mercury resonance
lines. This involves coating the inner surface of the lamp with
a thin-layer of crystalline phosphor. The phosphor is chosen
such that it will absorb the line emissions of the mercury vapor
and generate a broad band at a desired longer wavelength. Some
of the commerical filter fluorimeters use, for example, a 4-watt

low-pressure mercury-phosphor lamp having a broad emission band
with a maximum at 360 to 365 nm. Several mercury-phosphor lamps
are available. One has its output in the green region of the
spectrum with a band maximum at 525 nm, and is called a green
lamp. Another has its output in the blue region of the spectrum
with band maxima at 405 and 436 nm and is called a blue lamp.
A third kind of mercury-phosphor lamp is also available that has
a much wider band-output from 380 nm to 580 nm with peaks at
405, 436, 546, and 577 nm.

Xenon-Arc Lamps. A xenon-arc lamp is a short-arc gas-
discharge lamp with a quartz jacket, filled with xenon, at a
pressure of about 5 atmospheres at room temperature. A xenon-
arc lamp does not have as intense a light output as the mercury-
vapor lamp at its line emissions. However, the xenon-arc lamp
provides a good continuum from 250 to 800 nm, which is desirable
for the scanning type spectrometers. For this reason, most of
the modern spectrofluorimeters use a 150 watt xenon-arc lamp
as an excitation source. The main disadvantage of the xenon-arc
lamp is the fact that for a stable output from this lamp, an
expensive highly-regulated D.C. power supply is necessary. The
power supply directly affects the stability and life of the
lamp. As for any high-pressure short-arc lamps, the xenon-arc
lamp is also subject to the variation in the position of the
arc passing between the electrodes of the lamp. The variation
in the position of the arc is known as "arc-wander", and is
frequently a problem in fluorescence spectrophotometry because
it can cause a slight apparent shift in the excitation spectrum.
In modern 150 watt xenon-arc lamps the problem of arc-wander
has been virtually eliminated by better lamp construction and
by developing extremely stable power supplies. Some spectrofluorimeters have a special arrangement of entrance-slits to
minimize variations in the spectral distribution of the excitation light intensity due to small variations in the arc.
It should be noted that the spectral output of the xenon-arc
is qualitatively continuous. It does not generate all spectral
wavelengths with the same intensities. Rather, the production

of ultraviolet light of wavelengths shorter than 300 nm is much
less than the light of longer wavelengths.

Gas-discharge lamps usually employ high voltages and care
should be taken to avoid an electric shock. The medium to high-
pressure gas discharge lamps should be handled carefully to avoid
explosion. For example, a xenon-arc lamp which contains xenon
at a pressure of 5 atmospheres at room temperature, develops a
pressure of 20 atmospheres at operating temperature. While
transporting the lamps, the use of several layers of cloth or
other protective material to wrap the lamps is recommended.
Since gas-discharge lamps produce intense ultraviolet light,
one should never look at a lit gas-discharge lamp. The ultraviolet light can severely damage the retina of the eye.

The high-intensity ultraviolet light (180-220 nm) produced
by the gas-discharge lamps also converts molecular oxygen from
the surrounding air into ozone whose smell and physiological
effects can be irritating under normal lamp operating circumstances. The lamp housing must be constructed so as to prevent
the ultraviolet radiation from escaping into the room and should
be properly vented, usually by means of a tubular conduit,
running from the lamp housing to a hood or an exhaust system.
If the venting is impossible, the ozone can be catalytically
converted to molecular oxygen by passing the exhaust from the
lamp housing through a molecular sieve (10x) or platinum dust.
Commercial units are available which perform the latter de-
ozoning process.

Laser Excitation Sources. The development of ultraviolet and tunable lasers have brought new dimensions to fluorimetry. The nitrogen ultraviolet laser at 337 nm is a much more
intense source than the 366 nm line (triplet) of a mercury lamp.
Moreover, the former, due to its monochromicity, produces less
scatter. Tunable lasers in which a nitrogen ultraviolet laser
can be used as the primary excitation source can provide desired
wavelengths in the 360-650 nm range. The use of a tunable laser

as an excitation source would eliminate the need for excitation
filters or excitation monochromators.

Wavelength Selectors

It is possible to excite fluorescence and monitor the
emitted light without the use of any selecting device. However,
the possibility of exciting and measuring emission from extraneous species, as well as scattered incident radiation, along
with the emission from species of analytical interest, makes
the selection of limited regions of the spectrum for excitation
and for monitoring emitted light necessary.

Devices for limiting the range of wavelengths which excite
the analytical sample and the range of wavelengths emitted by
the sample which register on the detector system, fall into two
broad classes, filters and "monochromators". These differ
basically in the extent to which exciting and the emitted wavelengths are limited. Filters allow a relatively wide range of
wavelengths to excite the sample and to pass through to the
photodetector. Monochromators, which are usually diffraction
gratings with slit arrangements, inherently allow the passage
of a much smaller range of wavelengths which is ultimately determined by the optical characteristics of the diffraction
grating. However, it is possible to allow a wider range of
wavelengths to pass a monochromator by manipulation of the slits.
It should be noted that the more effective a monochromator or
filter is in achieving wavelength selectivity the less effective
it will be in allowing light flux to reach the sample or the
photodetector. Therefore, selectivity is usually achieved at
the expense of sensitivity. Although quartz prisms may theoretically be used in place of diffraction gratings as the dis-
persive elements in monochromators, they are not widely employed
because of their relatively high cost and inferior optical
properties. The greatest dispersion by the prism is in the
ultraviolet and not in the visible region where many emissions
occur. Quartz prisms with adequate dispersion properties

throughout the near ultraviolet and the visible region tend to
be excessively large.

Filters. An optical filter is optically homogeneous,
having desired characteristics of selective transmission, and
is used to transmit only a small part of the incident polychromatic radiation. Optical filters obey the Bouger-Lambert
law which requires that the spectral transmittance of two or
more optical filters used simultaneously must be equal to the
product of the spectral transmittances of each filter.

Optical filters can be made of tinted glass, gelatin
(Wratten filter) containing organic dyes usually sandwiched
between glass or lacquered for protection, a liquid solution of
absorbing substances, or can be interference filters. While
the interference filters depend on the constructive interference
of light rays for their transmission characteristics, the other
three filters selectively absorb unwanted portions of the incident polychromatic light and transmit only the light of desired
wavelengths. Since the absorption filters absorb radiation,
they can be heated and therefore should be cooled by a suitable
means. The absorption type filters fall into three general
categories; neutral tint, cutoff filters and bandpass filters.
Neutral tint or neutral density filters have a nearly constant
transmission over a wide range of wavelengths. They are used
to decrease the intensity of light uniformly and are used with
strongly fluorescing compounds. Cutoff filters have a sharp
cutoff in their transmittance characteristics, with complete
transmission on one side of the cutoff and little or no transmission on the other. These filters are used to cutoff stray
or unwanted light. They are especially useful to prevent the
scattered excitation light from falling on the fluorescence
detector. Bandpass filters are filters which transmit or reject
only a limited band of wavelengths. They are usually composite
filters constructed from sets of cutoff filters.

Interference filters are markedly different from absorption

type filters. The untransmitted radiant energy is eliminated
by destructive interference and reflection in an interference
filter. The interference filter consists of two highly reflective but partially transmitting films of silver, separated by
a spacer film of completely transparent material such as MgF_2.
Light of wavelength closest to the optical separation of the
silver films and integral multiples of that wavelength are
transmitted through the interference filter while other wavelengths are eliminated by destructive interference. Actually,
a narrow range of wavelengths (10-17 nm) centered about the
principal wavelength is transmitted through the interference
filter. Because little light energy is actually absorbed by
the filter, negligible heating of the filter occurs and this
type of filter is therefore particularly well-suited to use
with intense spectral sources.

The selection of a filter should be based on two criteria,
peak transmittance and bandpass width at the desired wavelength.
The transmittance should be high and the bandpass as narrow as
possible. A filter fluorimeter usually uses two filters; an
excitation or a primary filter to isolate the exciting light
and an emission or a secondary filter to isolate the emitted
light.

When choosing a primary filter, two things should be considered; the wavelength at which the sample absorbs maximally
(absorption maximum), and the relative intensity of the source
at that wavelength. Exciting at the absorption maximum will
give the highest sensitivity only if the continuum emission
source has a high output in the region of the spectral maximum.
If the source used is of the line emission type, i.e. mercury
lamp, the maximum sensitivity may be obtained by exciting the
sample at the strongest emission line (366 nm for the mercury
lamp). After a primary filter is selected, a secondary filter
can be chosen from a series of available "sharp cutoff" secondary filters so as to be compatible with the primary filter. A
secondary filter is considered "compatible" with a primary filter

if the short wavelength transmission of the former does not overlap significantly with the long wavelength transmission of the latter, thereby preventing scattered excitation radiation from reaching the detector. This is especially a problem in turbid or dust containing samples in which the occurrence of scattering of exciting light is high. Besides the consideration of compatibility, the filters should not possess any fluorescences of their own which cannot be excluded by the secondary filter. Gelatin filters, for example, fluoresce in the ultra-violet region.

Gratings. A grating disperses polychromatic light into its component wavelengths by utilizing the principle of destructive and constructive interferences of light. This is markedly different from the prism, which works on the principle of different refractive indices for different wavelengths. There are two kinds of gratings; the transmission grating and the reflection grating.

A transmission grating consists of a large number of parallel transparent lines and opaque lines arranged in alternate fashion. When a source of monochromatic light is incident on one side of a transmission grating, each transparent line acts like an independent line source of the parent radiation. Due to the wave nature of the radiation, interference will occur among the monochromatic light waves transmitted by the closely spaced transparent lines, resulting in reinforcement of the amplitude (intensity) at certain points and annihilation at others. This gives rise to a series of bright lines with dark regions between them. For a given grating the diffraction angle of the radiation transmitted depends on the wavelength of the radiation, and therefore polychromatic light, on passing through a grating is dispersed into its component wavelength.

For spectroscopic purposes, a reflection grating is used in preference to a transmission grating. The reflection grating is identical to the transmission grating, except for the fact

that the former uses a large number of grooves on a highly
reflective surface, such as aluminum, to cause the interference.

A small portion of a reflection grating along with the
critical parameters are shown in Figure 3.2. When the incident
radiation is perpendicular to the grating plane, the basic
equation that relates diffraction angle, Θ, of a wavelength λ,
to the distance between adjacent grooves, d, is

$$n\lambda = 2d \sin \Theta \qquad (3.1)$$

where n is the order of diffraction. When the angle made by
incident radiation with the normal of the grating is equal to
the angle of diffraction, the order of diffraction, n, is equal
to zero. The "zero order" diffraction corresponds to specular
reflection of the incident radiation. When n = 1, the diffraction is of the first order and it is the primary image of the
spectrum. When n = 2, the diffraction is of the second order
and it is the second image of the spectrum with the dispersion
twice that of the first order spectrum. The second order spectrum may superimpose itself on the first order spectrum, and
since a given wavelength in the former corresponds to twice
that value in the first order spectrum (i.e. the second order
250 nm spectral line will appear at 500 nm position of the
first order spectrum) this may cause analytical interference.
Therefore, a filter is usually used in association with a grating to filter out the higher order spectra.

The two most important factors that should be considered
in selecting a grating are the resolving power and the blaze
angle. The resolving power, R, directly depends upon the total
number of grooves in the grating and is defined as

$$R = \frac{\lambda}{\Delta\lambda} \ell N \qquad (3.2)$$

where $\Delta\lambda$ is the wavelength difference between two adjacent lines
that are barely distinguishable, λ is their average wavelength,

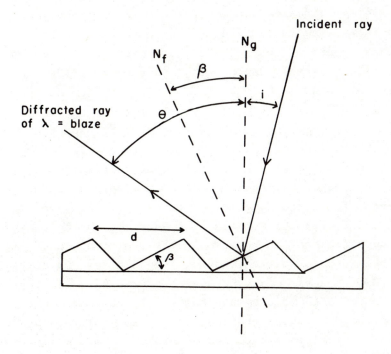

Fig. 3.2. Cross-section (exaggerated in size for clarity) of a
diffraction grating.

N_g = normal to grating;   Θ = angle of diffraction;

N_f = normal to groove face;   d = grating spacing in cm;

i = angle of incidence;   β = blaze angle.

The grating equation (when i ≠ Θ) = d (sin i + sin Θ).

ℓ is the length of the grating in centimeters, and N is the
number of lines or grooves per centimeter of the grating. The
higher the resolving power of a grating, the better the component wavelengths of polychromatic light will be separated from
each other. When the number of lines per unit length is increased, the dispersion in the first order of the grating increases, while an increase in the total number of lines, whether
achieved by increasing ℓ or N, increases resolution. Gratings
with resolving powers of up to 5×10^6 are available. Commercial instruments generally use gratings containing 15,000 or
30,000 lines per inch.

The blaze of a grating is the wavelength at which the
grating has the maximum efficiency. The grating efficiency at
a wavelength is the output of the grating at that wavelength
compared to that of an aluminum mirror at the same wavelength.
The efficiency of a grating decreases rapidly as the wavelengths
increasingly differ from the blaze; the decrease in the first-
order grating efficiency being more on the short wavelength
side of the blaze than on the long wavelength side. Usually,
the band-pass of the efficiency curve of a grating extends from
two thirds to twice the blaze wavelength. In luminescence
spectroscopy the excitation gratings are usually blazed at 300
nm and emission gratings are blazed at 500 nm, so that the former can be used optimally from 200 to 600 nm and the latter
from 335 to 1,000 nm.

Slits
    The slits, which aid in focusing the light onto the mono-
chromators and the detector regulate the range of wavelengths
which excite the sample and which ultimately pass onto the
detector. The slit width is the most important factor in determining the resolution of the instrument. The smaller the
slit widths employed, the narrower is the range of spectral
bandpass (bandwidth at one half the peak transmittance) and
the greater is the analytical selectivity. However, a smaller

slit width results in a decrease in the intensity of the
transmitted light and therefore results in a decrease in the
analytical sensitivity. In a grating instrument the bandpass
for a given slit is constant throughout the spectrum and depends
on the ruling of the grating. Slits are of two major types;
fixed and variable. Fixed slits are slots cut in an opaque
material. In an instrument using fixed slits, the desired slit
width can be obtained by using one of a series of fixed slits.
Fixed slits provide excellent reproducibility.

Variable slits are of two kinds; unilateral and bilateral.
The former contains two beveled blades, one of which is fixed
and the other can be moved with the help of a micrometer to
give continuous variation in slit widths within a limited range.
The bilateral slits are similar to the unilateral slits, except
that both the blades are moved in a symmetrical fashion. The
bilateral slits are preferred over the unilateral type because
the former maintains a constant center line while the center
line in the latter shifts as the slit width is altered. The
variable types of slits are considerably more expensive than
fixed slits but are more convenient for routine work. The
major disadvantage of variable slits is that with extensive use
the blade moving mechanism and the blade edges become worn,
resulting in loss of the slit calibration and reproducibility.
This is not a serious drawback in routine analytical work, but
in accurate spectroscopic studies of physicochemical phenomena
the precise knowledge of the excitation and emission slit widths
may be important.

Sample Compartments and Sample Cells

Sample compartments contain the cells in which the analytical samples are excited. They are usually painted with flat
black paint, to minimize stray exciting light, and are covered
when the instrument is in operation, to exclude external light.
In most instruments the sample compartment is arranged so that
the fluorescent light, emitted by the sample, exits from the

compartment to the emission monochromator at right angles to the line of entry of the exciting light. This is commonly called a 90° configuration. Any other configuration of cell compartment may be used, but the minimum interference from stray exciting light is obtained with a 90° configuration. If the sample has strong absorption at the wavelength of excitation, concentration quenching may be a problem with a 90° configuration, especially when the sample has a small fluorescence quantum yield. In this case a front-surface configuration with a solid-sample accessory or a microcell can be used to eliminate the problem.

The sample cells employed in ambient temperature fluorescence spectroscopy are usually 1 cm^2, and are made of pyrex glass, high quality quartz or fused silica. Pyrex cells are useful for measurements involving excitation above 320 nm and are inexpensive compared to quartz or silica cells. For measurements involving ultraviolet radiation, quartz or silica cells are necessary. Fused silica (Suprasil) cells are usually superior because they demonstrate less background fluorescence than their counterparts made of quartz.

Detectors

The detectors employed in ultraviolet-visible emission spectroscopy consist of a photomultiplier tube. A photomultiplier is a phototube capable of effecting photocathode emission witn multiple cascade stages of electron amplification to achieve a large and linear amplification of primary photocurrent within the tube itself. The electric current produced by a photomultiplier, upon exposure to light, is then amplified by an amplifier to a measurable level.

Photomultiplier tubes do not respond equally to all wavelengths of light. As a result, the electrical signal put out by a given photomultiplier does not faithfully represent the relative intensity of the wavelength spectrum it receives but,

rather, is biased toward certain wavelength regions of the
visible and ultraviolet spectrum and therefore causes spectral
distortion. Most photomultiplier tubes used in commercially
available instruments have maximum spectral response, from about
300 to 500 nm and are suitable for the great majority of fluorescence and phosphorescence work. However, in the long wavelength region of the spectrum (wavelength greater than 550 nm)
the electrical response of these phototubes falls off dramatically, resulting in poor analytical sensitivity. If spectral
measurements are to be performed frequently at long wavelengths
it is often desirable to employ a special red sensitive phototube.

Readout System

   Readout systems for fluorescence spectroscopy may be
either galvanometers, recorders or oscilloscopes. Galvanometers, the meters supplied with the spectrofluorimeters are
convenient for analytical measurements at fixed wavelength but
are cumbersome for spectral work, as the spectrum must be plotted manually from point by point readings on the galvanometer.
Recorders provide a permanent record of the fluorescence excitation or emission spectrum and are of two basic types, x-y
and x-t. The x-t or strip chart recorder displays fluorescence intensity on the x axis and wavelength on the t (time)
axis. The t axis is traversed at a constant rate of speed and
may be slaved to the monochromator drive motors which, with the
x-t recorders, are also operated at constant scanning rate.
Several constant scanning speeds are available on most x-t
recorders allowing spectra to be expanded or compressed to the
desired scale. x-y recorders are usually about twice as expensive as x-t recorders. As in the x-t recorder, the x-axis
corresponds to fluorescence intensity which is represented by
the deflection of a pen driven by the output from the photodetector. However, the y axis of the x-y recorder is continuously synchronized with the scanning speed of the monochromators. Thus, the slowing down of the monochromator scanning

rate in areas of spectroscopic interest, or the speeding up of
the monochromator scanning rate in spectroscopically uninterest-
ing areas is continuously matched by the scanning rate of the
y-axis. This allows an undistorted spectrum to be taken even if
different portions of the spectrum were scanned at different
rates. Moreover, a spectrum can be made to retrace itself by
back-scanning with an x-y recorder. With an x-t recorder the
rewinding of the chart is not synchronized with the back-rota-
tion of the monochromators so that the retracing capability is
absent. Perhaps the greatest limitation of recorders, in
general, is the rather long response time of the recorder pen
(0.1 - 0.5 sec.). This limits the rate of which an accurate
spectrum can be swept. The oscilloscope circumvents the problem of pen response time by tracing out the spectrum electronically rather than mechanically. However, good oscilloscopes
may be more expensive than recorders, and in order to obtain a
permanent record of the displayed spectrum, some rather elegant
photographic apparatus may be required.

Instrumental Distortion of Spectra

Because of the wavelength variable output of the lamp and
responses of the monochromators and phototube, fluorescence
excitation and emission spectra taken on conventional spectrofluorimeters are not "true" excitation and emission spectra
(properties of the analyte alone), but rather, are apparent
excitation and emission spectra distorted by instrumental
response. For routine quantitative analysis this is usually
unimportant. However, the "true" fluorescence excitation spectrum is identical to the absorption spectrum of the analyte.
The ability to take true excitation spectra is thus, analytically, very desirable because it facilitates identification of
organic molecules at concentrations far lower than possible by
most other means. The "true" fluorescence spectrum is necessary
for the accurate calculation of fluorescence quantum yields
(from Equation 3.7). Moreover, because the distortion of the
apparent spectrum depends upon the manufacturer of the

spectrofluorimeter components as well as upon the age of the
lamp and phototube, "true" excitation and emission spectra are
highly desirable in order to compare results taken on different
instruments. Apparent excitation and emission spectra can be
corrected for instrumental response to yield "true" excitation
and emission spectra by means of commercially available correction accessories which vary from manufacturer to manufacturer in
mode of operation.

Low Temperature Fluorescence

Low temperature fluorescence studies may be carried out by
using a small Dewar flask with a quartz window or nipple at its
bottom (Figure 3.3). The Dewar is placed in the sample compartment
of a fluorimeter such that the quartz nipple is aligned with the
excitation and emission optics. The Dewar may be filled with
liquid nitrogen or any other cryogenic fluid. The sample cell,
in the form of a thin-walled small-diameter glass or quartz tube,
containing the sample, is then inserted into the filled Dewar
flask so that the sample is exposed to the exciting light through
the quartz nipple. Fluorescent light also passes out of the
sample to the detector through the quartz nipple. A vacuum may
be maintained in the sample compartment to avoid light scattering
due to condensation of water vapor from the air on the nipple.
Alternatively, dry nitrogen can be circulated in the sample compartment to achieve the same result. The solvents used for low-
temperature fluorescence should have the property of solidifying
to a clear, rigid "glass" at low-temperatures. Solvents used in
phosphorimetry (see Table 3.1) can be employed for low-temperature
fluorescence. To obtain low-temperature fluorescence in an
aqueous solution, a 5-10% aqueous methanol can be used as a solvent, as pure water cracks when frozen, resulting in light
scattering from cracked surfaces (1). Light scattering may also
be caused by bubbling of the liquid nitrogen filling the Dewar.
This may be eliminated by the addition to the liquid nitrogen a
small amount of liquid helium.

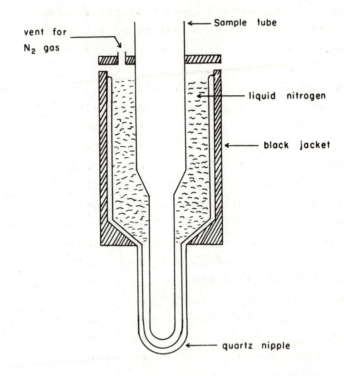

Fig. 3.3.   Schematic diagram of Dewar flask for low temperature
            fluorescence or phosphorescence studies.

Phosphorescence Measurements

Phosphorescence is usually observed at low temperature. If a sample phosphoresces but does not fluoresce, the phosphorescence measurements can be done with the set-up used for the low-temperature fluorescence measurement. However, if a sample fluoresces and phosphoresces, the phosphorescence measurements are made, based upon time-resolution of the slow-decaying phosphorescence relative to the fast-decaying fluorescence. A typical phosphorimeter is essentially a fluorimeter (filter or grating type) in low temperature mode with a device called a mechanical phosphoroscope which allows only the long-lived phosphorescence to reach the detector. The widely used rotating can phosphoroscope (Figure 3.4) is a mechanical chopper, consisting of a can with lateral apertures 180° apart, which is inserted into the sample compartment in such a way as to enclose the Dewar flask and its contents. The quartz nipple of the flask is aligned with the apertures in the side of the phosphoroscope. The phosphoroscope can be rotated at several thousand revolutions per minute with a variable speed motor. Because the apertures in the phosphoroscope are 180° apart and the excitation-emission optics are in 90° configuration, when one aperture is facing the excitation optics the sample is excited but the emission optics are shielded from the light emitted by the sample. Hence no light falls on the detector. When the can rotates through 90° the exciting light is cut off from the sample so that excitation ceases. The time the can takes to make this quarter revolution is very long compared to the mean lifetimes of the fluorescing species in the sample, but is generally short compared to the mean lifetimes of the phosphorescing species. Hence, when the aperture of the can faces the detector optics, all fluorescence initially excited has died out and the only light reaching the detector is that of the long-lived phosphorescence.

For converting a fluorimeter with a straight line or 180° configuration to a phosphorimeter, a mechanical phosphoroscope called a Becquerel disc phosphoroscope (Figure 3.5) is used. This phosphoroscope consists of two discs with notches cut in them at

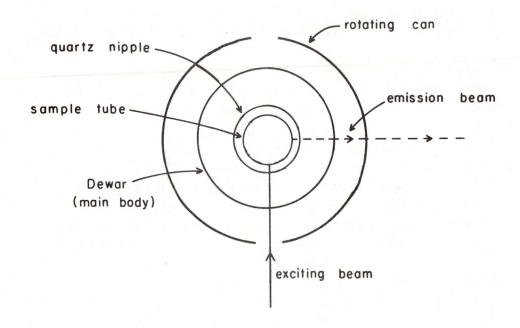

Fig. 3.4. Top view of sample compartment for phosphorescence studies with rotating can phosphoroscope.

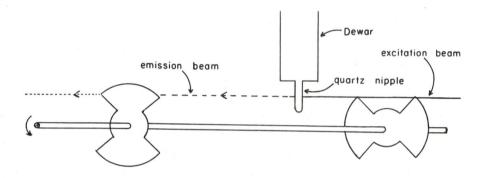

Fig. 3.5. Schematic diagram of a Becquerel disk phosphoroscope.

equal intervals and are mounted a few centimeters apart, on a
common axis, in a staggered configuration. The phosphorescence
is positioned so as to accommodate the sample in between the two
discs with its axis parallel to the optical path. The discs are
rotated by a motor and as in the case of the can-type phosphoroscope, excitation occurs out of phase with the detection of the
emitted light.

Phosphoroscopes are characterized by the observation-efficiency factor, α', the time required for one cycle of excitation and observation, τ_c, the delay-time or the time between
termination of excitation and initiation of observation during
one cycle, t_D, and the exposure time or the duration of excitation of the sample during one cycle, t_E.

The observation-efficiency factor, α', is defined as the
ratio of the phosphorescence signal obtained with a phosphoroscope to the phosphorescence signal observed in absence of the
phosphoroscope. If the phosphorescence decay-time of a sample,
τ, is large compared to the cycle time, t_c, the value of α' will
be a constant for any phosphoroscope speed. However, if $\tau < t_c$,
some of the phosphorescence during the time of observation will
be significant. This will result in a lower α'. In this case,
the value of α' and hence the sensitivity can be increased by
increasing the rate of modulation. This can be achieved by
increasing the motor speed and/or increasing the number of
openings in the can or disc. O'Haver and Winefordner (2) have
shown that with a can-type phosphoroscope at a maximum speed of
15,000 revolutions per minute, only the emissions having τ of
greater than one millisecond can be measured. From the analytical point of view, it is important to determine whether or
not α' is constant with phosphoroscope speed before performing
quantitative analysis with a phosphorimeter.

Sample Cells and Solvents for Phosphorimetry
The cells used for phosphorescence are the same as those

used for low-temperature fluorescence. The solvents used for
phosphorescence measurements must have sufficient viscosity at
the cryogenic temperature to permit a high yield of phosphorescence or should have the property of solidifying to a clear rigid
"glass" at low temperatures. Table 3.1 lists some of the solvents
and mixture of solvents used in phosphorescence measurements.
The most commonly used solvent for phosphorescence measurements
is a mixed solvent containing diethyl ether, isopentane and ethyl
alcohol in a 5:5:2 (v/v/v) ratio, and is popularly known as EPA.
In selecting a solvent for phosphorescence measurements the
following criteria should be considered; solubility of the solute, and viscosity and optical properties of the solution at the
desired temperature. The small diameter of the cells and inhomo-
geneities of the cell and sample solutions used in phosphorimetry
require precise positioning and repositioning of the cell with
respect to the optical arrangement. This problem can be overcome
by using a rotating sample cell, which randomizes the rigid solvent cracks and cell alignments (3,4). Although it is desirable
to have the solvent form a clear glass in phosphorimetry,
Zweidinger and Winefordner (4) in a recent report, have shown
that "snows" or highly cracked glasses of solutions completely
randomize the light scatter. A 10-30% aqueous methanol and
several hydrocarbons when used as solvents for low-temperature
phosphorescence measurement give excellent "snows". The random-
ization of light scatter by the "snows" can be further enhanced
by using the rotating cell holder.

Measurement of Fluorescence Quantum Yields

The fluorescence quantum yield, ϕ_f, of a molecular species
may be determined experimentally by comparing the integrated
fluorescence intensity, I_f, (the area under the corrected
fluorescence spectrum) and the absorbance at the wavelength of
excitation to the corresponding quantities belonging to a
solution of a reference compound of known quantum yield. Quinine
bisulfate in 0.1 NH_2SO_4, has a quantum yield of 0.55 and is the
standard most often employed. According to equation (1.12) for

very dilute solutions of quinine (Q) and of the compound whose
quantum yield is to be determined (U), under identical conditions
of excitation,

$$\phi_{fu} = \frac{I_{fu}}{I_{fQ}} \frac{A_Q}{A_u} \phi_{fQ} \tag{3.3}$$

where A_Q and A_u are the absorbances of the test compound and of
quinine bisulfate, respectively (A = $\epsilon c \ell$). Reported quantum
yield values typically fall between 0.1 and 1.0 for analytically
useful determinations.

Measurement of Luminescence Lifetime

The decay-time of the fluorescence of organic molecules is
generally of the order of several nanoseconds, and so techniques
for its measurements are generally rather elegant and require
expensive instrumentation. These include the stroboscopic
(pulsed-flash) technique and the phase-shift technique. The
stroboscopic method uses special high speed flash-lamps, capable
of generating sub-nanosecond light pulse. Commerical flash-
lamp systems of this type are available (TRW Model 31A Nanosecond Spectral Source, TRW Instruments, El Segundo, California;
and the Xenon Model 437 Nanopulser and 783A Nanopulse Lamp,
Xenon Corp., Medford, Mass.).

The TRW Model 75A Decay-Time Fluorimeter uses the TRW
Model 31A Nanosecond Spectral Source containing either a nitrogen or a deuterium flash-lamp. The lamp is made to excite the
sample at a rate of 2 or 5 KHz and the resulting fluorescence
signal (fluorescence appearance and decay) is detected by a
photomultiplier tube and displayed on one channel of a dual-
beam oscilloscope as a function of time. The second channel of
the oscilloscope displays a simulated decay-time curve from a
minicomputer assembly (TRW Model 32A Decay Time Computer) that
must be standardized with the lamp-pulse curve prior to fluorescence lifetime measurement. To obtain the fluorescence

lifetime, τ, of a sample, the computer curve is manipulated to
exactly match the fluorescence decay curve of the sample and the
value of τ directly read from the computer readout. This technique is limited by the matching of the computer-simulated decay
curve to the actual fluorescence decay, and usually gives reli-
able values of τ for simple exponential decay.

Another instrument available for lifetime measurements is
the ORTEC Model 9200 Nanosecond Fluorescence Spectrometer (ORTEC
Incorp., Oak Ridge, Tenn.). This instrument is based on the
recently developed "single-photon counting" technique which is
acknowledged to be the best method of measuring very low light
levels. The spectrometer uses a nanosecond light pulser with
full width at half maximum of 2 nsec. The typical flash rates
used are of the order 20,000 to 40,000 pulses/sec. Photomultiplier tubes are used for photon counting and the data are stored
in a multi channel representing a time period after the lamp
flash. This system gives better time resolution, high sensitivity and a digital output which facilitates computer processing
of the fluorescence decay curves.

The phase-shift technique for determining fluorescence
lifetime involves high frequency modulation of excitation and
fluorescence light, both at the same frequency. The high frequency modulation of light is achieved by using electronic
shutters (e.g. Keir cells), electro-optical devices, or diffraction of light as it travels through a medium vibrating at
ultrasonic frequencies. Although the excitation and emission
are modulated at the same frequency, the observed fluorescence
will lag behind the excitation by a period depending upon the
fluorescence decay-time. This lag is detected as an electrical
phase-shift. For simple exponential decays the fluorescence
lifetime, τ, is related to the phase-shift, θ, by the following
relationship,

$$\tau = \frac{\tan \theta}{2 \pi \nu} \qquad (3.4)$$

where ν is the frequency of modulation. Phase shifts of 10^{-3}
degrees or less can be accurately measured yielding accurate
values of τ.

Measurement of Concentration

The measurement of analyte concentration in fluorescence
spectroscopy may be accomplished by several distinct experimental
methods. The simplest of these, applicable only to intrinsically
fluorescent species, is the direct method, which involves the
measurement of the fluorescence intensity of the analyte itself,
either with or without prior separation from interfering substances. A nonfluorescent or weakly fluorescent substance may
be converted into a form more suitable for fluorescence analysis
by means of appropriate chemical reactions. Such methods are
called chemical methods. If the substance to be determined is
nonfluorescent and yet possesses the ability to quench the
luminescence of some luminescent compound, then the substance
may be determined indirectly by the measurement of the reduction
in luminescence intensity of the luminescent compound. Such a
process would constitute a quenching method. Finally, there is
the possibility of energy transfer methods, which involve the
absorption of excitation light by a donor species, transfer of
the energy from the donor to an acceptor species, which may be
either another molecule (intermolecular) or another portion of
the same molecule (intramolecular).

All of the above methods are relative methods, and as such,
require some sort of calibration procedure. The most common
calibration method involves the preparation of a series of
standard solutions (serial dilutions) of the analyte. These
standards are then treated as the unknown samples themselves,
according to the particular measurement method used. Thus, in
the direct method, the fluorescence intensity of each standard
is measured directly. In the chemical method, appropriate
reagents are added to each standard, and the mixture is allowed
to react before the fluorescence intensity is measured. In the

intermolecular energy transfer methods, a suitable acceptor or donor species must be added. The relative fluorescence intensity of each of the standard solutions prepared in the above way is measured, blank-corrected, and plotted _vs_ the analyte concentration, usually on log-log coordinates. Luminescence analytical curves are typically linear over several decades of concentration, have log-log slopes near unity, and exhibit a plateau region at high analyte concentrations.

Unknown sample solutions are treated and measured in exactly the same way as the standards, and their concentrations are interpolated from the analytical curve. Linear analytical curves are highly desirable, because interpolation from a linear plot is relatively easy.

Only in the most ideal analysis will all of the measured light emission from the sample be due to the analyte. The purpose of the blank solution is to correct for absorbing and fluorescing species other than those related to analyte concentration.

All reading of the luminescence signals of samples and standards must be blank-corrected by subtracting the signal reading given by the blank solutions. Several types of blank solutions are used in practice. The ideal or true blank would in principal contain everything contained in the unknown samples in the same concentrations as in the unknown samples except the analyte. A true blank is seldom possible for real analyses in complicated systems. Thus, some approximation to a true blank is usually made. The simplest is a solvent blank, consisting simply of the solvent used to make up the standards and dissolve and dilute the samples. Such a blank would correct only for fluorescent impurities in the solvent and would be suitable only in the direct measurement of a fluorescent compound uncontaminated by fluorescent impurities. More satisfactory is a reagent blank which contains, in addition to the solvent, each of the various reagents in the same concentrations as used in the

treatment of samples and standards. A reagent blank is useful
in the chemical, quenching, and energy transfer methods, as it
corrects for absorption by and luminescence of impurities in the
added reagents. Neither a solvent blank or a reagent blank,
however, corrects for the absorption by and luminescence of
contaminants and matrix substances originally in the sample
itself. If these interferences cannot be distinguished spec-
trally, chemical or physicochemical separation may be necessary.

If a separation step is impossible or undesirable, it may
be possible to compensate for the presence of interfering substances by the preparation of an internal blank which is very
nearly a true blank. An internal blank is produced by adding to
the sample, after its fluorescence intensity has been measured,
a nonfluorescent compound which specifically reacts with the
analyte to yield products which are not fluorescent at the excitation and emission wavelengths used for the analysis. Alternatively, a compound which specifically quenches the analyte
luminescence may be used. If the instrumental sensitivity should
change between the time the analytical curve is measured and the
time that the sample is measured, (e.g. aging of the excitation
lamp, fatigue of the multiplier phototube, etc.), then it is
possible to correct for this by use of a reference solution.
The reference solution should contain a very stable, easily
purified fluorescent species which absorbs and fluoresces in the
approximate wavelength range of interest (e.g. a dilute solution
of quinine bisulfate in 0.1M H_2SO_4). The concentration of the
reference species should be on the linear portion of the curve.
After an analytical curve is determined or during the establishment of the analytical curve, the fluorescence intensity of the
reference solution is also measured. When unknown samples are
to be fluorometrically measured at some later time, the instrumental sensitivity is adjusted to give the same reading originally obtained for the reference solution. This adjustment is best
accomplished by varying the amplifier gain or phototube voltage
rather than the monochromomator spectral bandwidth, as it is the
former parameters which alter instrumental sensitivity by their

variance with time.

Time-Resolved Fluorimetry (8)

In analytical fluorimetry, as it is currently practiced, selectivity, the resolution of the fluorescence of a substance of analytical interest from the emissions of interfering substances, is achieved either by manipulation of the chemistry of the sample or by resolution of the excitation or emission spectra of the analyte and interferences by means of monochromators. In many cases the analytical selectivity achieved by these means is inadequate for analytical purposes and time-consuming chemical separations are then required prior to the measurement of fluorescence. However, in many cases it is possible, in the event of several overlapping fluorescences, to achieve spectral resolution by means of the measurement of the differing decay-time characteristics of the overlapping fluorescences. This is accomplished by exciting the sample with a pulsed excitation source whose rise and decay-time is shorter than the lifetime of the lowest excited singlet state of the analyte of interest. The fluorescence observed from the sample will then show the composite decay characteristics of all fluorescing species whose decay times are longer than that of the pulsed source. If the decay characteristics of the pulsed source and those of the sample are, respectively, fed into the two channels of a dual beam oscilloscope, the decay curve from the source can be used to calibrate the time-axis of the oscilloscope screen while the decay curve from the sample can be taken as representative of the decay characteristics of the sample. If the decay-times of the fluorescing components of the sample are different (even though their fluorescence spectra may overlap), a plot of the logarithm of the intensity of fluorescence of the sample (calibrated against suitable standards) against the decay-time, will yield several straight line segments. This plot is similar to that employed in radiochemical analysis for the resolution of nuclides of differing lifetimes. Each straight line segment corresponds to the decay of a different fluorescing component of

the sample, the ones demonstrating the steepest slopes having
the shortest decay times. The extrapolation of each line segment
to the intensity (concentration) axis will yield the concentration
of the corresponding fluorescing species. The identity of each
species can be established by comparing the time axis intercept
of each line segment with the decay-time characteristics of the
fluorescence of a pure sample of each analyte. In this way, the
concentrations of several species, whose fluorescences overlap,
can be determined without chemical separation. The principal
difficulties of this method lie with the currently primitive
instrumentation available for lifetime measurement and the need
for computerization in cases where several fluorescing materials
of only slightly differing lifetime of fluorescence are present.

Selective-Modulation Fluorimetry (9)

A recently established method alternative to time-resolution
for the determination of fluorescing analytes whose excitation
or emission spectra overlap is selective modulation fluorimetry.
In this method, either the excitation wavelength is modulated
(scanned rapidly back and forth over a small wavelength interval)
and the fluorescence spectrum _in_ _toto_, or the fluorescence wavelength is modulated and the excitation monochromator scanned.
This technique produces essentially a difference fluorescence or
excitation spectrum which by means of frequency-selective electronics (a lock-in amplifier) and judicious choice of the
modulation interval, enables the emissions of the fluorescing
components to be measured independently. Presumably, future
demand for this instrumentation will bring the cost of the
sophisticated electronics down to the point where it can enjoy
wide application in analytical laboratories.

Microspectrofluorimetry (8)

Fluorescence microscopy has been employed for some time in
clinical chemistry for qualitative immunofluorescent identification of pathogens and in the identification of cellular structures

by fluorescent staining techniques. However, recently the
development and commercial availability of monochromated microspectrofluorimeters capable of quantitative measurement by means
of phototubes has become a reality.

The microspectrofluorimeter is, in essence, a fluorescence
spectrometer coupled to a dark-field microscope. The microscope
stage serves as the sample compartment and is flooded by exciting
light coming through the excitation monochromator of the spectrometer. The fluorescence from a specimen on the microscope stage
is focused through the optics of the microscope and may be observed visually through the ocular or passed through an emission
monochromator to a photodetector and on to a suitable readout
device. The dark-field optical arrangement of the microscope
prevents the exciting light which floods the microscope stage
from being transmitted through the barrel of the microscope and
thereby eliminates this intense light as a source of interference. Modern microspectrofluorimeters can be employed to take
excitation and emission spectra of cross sections as small as 1
micron in diameter. Consequently, this device is useful, not
only for the fluorimetric analysis of small tissue samples but
also for the quantitation of fluorogens in specific parts of
single cells. However, because of the glass optics employed in
reasonably priced optical microscopes and the blue fluorescent
background from cellular constituents, microspectrofluorimetry
is practically limited to the study of fluorochromes which
luminesce at wavelengths greater than 450 - 500 nm.

TABLE 3.1

Some Solvent Systems for Phosphorimetric Analysis*

Solvent System	Composition (V/V)	Nature	Temperature, $^\circ$K
ONE COMPONENT			
Sulfuric acid	98%	acidic	At least to 210
Phosphoric acid	88%	acidic	At least to 210
Acetic acid	glacial	acidic	77,90
Boric acid	100%		Room temp.
Ethyl alcohol	> 95%	alcoholic	77
n-Propanol	100%	alcoholic	77
Glycerol	100%	alcoholic	183-195
Propylene glycol	100%	alcoholic	183
Triethanolamine	100%	alcoholic	193-213
Glucose	100%	carbohydrate	Room temp.
Sucrose	100%	carbohydrate	Room temp.
Diethyl ether	100%	ether	77
2-Methyltetrahydro-furan	100%	ether	77
Bromoform	100%	halide (organic)	77,90
2-Bromobutane	100%	halide (organic)	77
Chloroform	100%	halide (organic)	77,90
Methane	100%	hydrocarbon	4.2
Isopentane	100%	hydrocarbon	77
3-Methylpentane	100%	hydrocarbon	77
Pentane(technical grade, 1:1, n-pentane: isopentane)	100%	hydrocarbon	77

Solvent System	Composition (V/V)	Nature	Temperature, $^{\circ}$K
Petroleum ether (58-60°C fraction)	100%	hydrocarbon	77
Pentene-2 (cis)-pentene-2 (trans) (mixed isomers)	100%	hydrocarbon	77
Paraffin oil (Nujol)	100%	hydrocarbon	183-195
Cellulose acetate	100%	polymeric	77-300
Cellophane	100%	polymeric	77-300
Lucite (polymethyl-methacrylate)	100%	polymeric	80-300
Plexiglass	100%	polymeric	Room temp.
Polyacrylonitride; Orlon A	100%	polymeric	77-300
Polyvinyl alcohol	100%	polymeric	Room temp.
TWO COMPONENTS			
Acetic anhydride-phosphoric acid	17:25	acidic	At least to 210
Ethyl alcohol-concentrated hydrochloric acid	19:1	alcoholic	77
Ethyl alcohol-Methyl alcohol	4:1 to 5:2	alcoholic	77
Isopropanol-isopentane	3:7, 2:8	alcoholic	77
n-Butanol-isopentane	3:7	alcoholic	77
n-Propanol-isopentane	2:8	alcoholic	77
Diethyl ether-ethyl alcohol (96%)	1:2	alcoholic	77
Ethyl alcohol-glycerol	11:1	alcoholic	77

Solvent System	Composition (V/V)	Nature	Temperature, $^{\circ}$K
n-Propanol-diethyl ether	2:5	alcoholic	77
n-Butanol-diethyl ether	2:5	alcoholic	77
Diethyl ether-isopropanol	3:1	alcoholic	77
Water-propylene glycol	1:1	aqueous	183-195
Water-ethylene glycol	1:2	aqueous	123-150
Phosphate buffer (0.01 M, pH = 7)-propane-1,2-diol	1:1	aqueous	77
Ethyl alcohol-ammonia (28% aqueous)	< 20:1	basic	77
Ethyl alcohol-sodium hydroxide (0.5% aqueous)	< 20:1	basic	77
Methylhydrazine-methylamine	2:4	basic	77
Di-n-propyl ether-isopentane	3:1	ether	77
Di-n-propyl ether-methylcyclohexane	3:1	ether	77
Di-n-propyl ether-n-pentane	2:1	ether	77
Diethyl ether-pentene-2 (cis)-pentene-2 (trans)	2:1	ether	77
Diethyl ether-isopentane	1:1 to 1:2	ether	77
n-Pentane-n-heptane	1:1	hydrocarbon	77
Methylcyclohexane-n-pentane	4:1 to 3:2	hydrocarbon	77

Solvent System	Composition (V/V)	Nature	Temperature, ^{o}K
Methylcyclohexane isopentane	4:1 to 1:5	hydrocarbon	77
Methylcyclohexane-methylcyclopentane	1:1	hydrocarbon	77
Cyclohexane-decalin	1:3	hydrocarbon	20-77

THREE COMPONENTS

Trifluoroacetic acid-methylamine-trimethylamine	4:11:5	acidic	77
Ethyl alcohol-isopentane-diethyl ether (EPA)	2:5:5	alcoholic	77
Isopropanol-isopentane-diethylether	2:5:5	alcoholic	77
Ethyl alcohol-methyl alcohol-diethyl ether	8:2:1	alcoholic	77
Diethyl ether-isooctane isopropanol	3:3:1	alcoholic	77
Diethyl ether-isooctane-ethyl-alcohol	3:3:1	alcoholic	77
Isopropanol-methylcyclohexane-isooctane	1:3:3	alcoholic	77
Ethyl cellosolve-n-butanol-n-pentane	1:2:10	alcoholic	77
Triethylamine-diethyl ether-isopentane	3:1:3	basic	77
Triethylamine-diethyl ether-n-pentane	2:5:5	basic	77

Solvent System	Composition (V/V)	Nature	Temperature, °K
Triethylamine-iso-pentane-diethyl ether	2:5:5	basic	77
Diethyl ether-ethyl alcohol-ammonia (28% aqueous)	10:9:1	basic	77
n-Butyl ether-Isopropyl ether-diethylether	3:5:12	ether	77
Ethyl iodide-isopentane-di-ethyl ether	1:2:1	halide (organic)	77
Ethyl alcohol-methyl alcohol-ethyl iodide	16:4:1	halide (organic)	77
Ethyl alcohol-methyl alcohol-propyl iodide	16:4:1	halide (organic)	77
Ethyl alcohol-methyl alcohol-propyl chloride	16:4:1	halide (organic)	77
Ethyl alcohol-methyl alcohol-propyl bromide	16:4:1	halide (organic)	77
Ethyl iodide-ethyl alcohol-methyl alcohol	5:16:4	halide (organic)	77
Ethyl alcohol-isopentane - diethyl ether and 0.01 M in dry HCl	1:1:1	halide (organic)	77

Solvent System	Composition (V/V)	Nature	Temperature, $^{\circ}$K
FOUR COMPONENTS			
Diethyl ether-isopentane-dimethylformamide-ethyl alcohol	12:10:6:1	ether	77
Diethyl ether-isopentane-dimethylformamide-ethyl alcohol	2:4:1:16	ether	77
Ethyl alcohol-isopentane-diethyl ether-chloroform (EPA-CHCℓ_3, 12:1)	24:60:60:1	halide (organic)	77
Ethyl bromide-methylcyclohexane-isopentane-methylcyclopentane	1:4:7:7	halide (organic)	77

* Data taken from references 6 and 7.

REFERENCES

1. Lukasiewicz, R. J., Mousa, J. J. and Winefordner, J. D., Anal. Chem. 44, 963 (1972).

2. O'Haver, T. C. and Winefordner, J. D., Anal. Chem. 38, 682 (1966).

3. Hollifield, H. C. and Winefordner, J. D., Anal. Chem. 40, 1759 (1968).

4. Zweidinger, R. A. and Winefordner, J. D., Anal. Chem. 42, 639 (1970).

5. Parker, C. A. and Rees, W. T., Analyst 85, 587 (1960).

6. Winefordner, J. D., St. John, P. A. and McCarthy, W. J., "Phosphorimetry as a Means of Chemical Analysis", Chapter II in Fluorescence Assay in Biology and Medicine, Volume II, S. Udenfriend, Ed., Academic Press, New York, N.Y., 1962.

7. Guilbault, G. G., Practical Fluorescence - Theory, Methods and Techniques, Marcel Dekker, New York, N.Y., 1973.

8. A. A. Thaer and M. Sernetz, Eds., Fluorescence Techniques in Cell Biology, Springer-Verlag, New York (1973).

9. O'Haver, T. C. and Parks, W. M., Anal. Chem. 46, 1886 (1974).

CHAPTER 4
APPLICATIONS

Organic Compounds

A number of organic compounds possess native fluorescence
which can be used for their qualitative and quantitative analysis. DeMent (1) in his book, <u>Fluorochemistry</u>, lists over 2800
compounds that have native fluorescence in the uv or visible
region of the spectrum. The native fluorescence is limited to
those organic compounds containing aromatic rings or highly
conjugated aliphatic systems such as carotene. Even in these
classes of compounds not all members fluoresce. However, com-
pensating for the lack of universality of the directly excited
fluorescence is the specificity and great sensitivity of analysis resulting from the fluorescence properties inherent to the
analyte of interest.

Those aromatic and highly conjugated aliphatic compounds
which do not show native fluorescence in aqueous solution may
become fluorescent if the chemical parameters of the analytical
medium are changed. A change of solvent may make a non-fluorescent compound fluorescent. For example, 8-anilinonaphthalene-
1-sulfonate ion which has weak fluorescence in aqueous solution
becomes intensely fluorescent in non-polar solvents like alco-
hols. Aromatic aldehydes, such as pyrene-3-aldehyde, naph-
thalene-2-aldehyde, anthracene-9-aldehyde, and acenaphthene-3-
aldehyde fluoresce in hydroxylic solvents but do not fluoresce in hydrocarbon solvents (2). In the case of compounds
containing acidic or basic groups, a change in pH of the aqueous
solution may induce fluorescence or phosphorescence either as
a result of the protonation or ionization of these groups. For
example, benzoate anion which is non-fluorescent in aqueous
solution at pH 7 becomes fluorescent on protonation in strongly
acidic solution (3). Aniline fluoresces intensely in neutral
form at pH > 7 while its protonated form at acidic pH is

non-fluorescent (4).

When an analyte is to be determined by fluorimetry or
phosphorimetry in a solution which also contains interfering
luminescing impurities, separation of the analyte by extraction
or chromatography may be necessary before the fluorescence assay
of the desired components can be carried out. This is particularly important for samples of biological origin, as many
compounds commonly present in biological media are fluorescent
and therefore are potential interferences in the fluorimetric
assays. Separation of the analyte from interfering luminescing
impurities may not be necessary if i) the concentrations of the
interferences are sufficiently low as not to absorb the exciting light, ii) if the emissions from the interfering species
are at wavelengths other than the analytical wavelength chosen
for the analyte, and iii) if the interferences do not add or
subtract from the fluorescence by energy exchange with the
excited analyte molecules.

If the analyte and the impurities fluoresce in the same
spectral region and if the spectral resolution is inadequate
for the analysis of the analyte, it is often possible to achieve
sufficient instrumental resolution by varying chemical parameters of the analytical medium, such as pH. For example, if
the interfering impurities contain acidic or basic fundamental
groups not present in the analyte, selective shifting of the
excitation and emission spectra, or selective quenching of the
fluorescence of interfering species may be affected by ionization or protonation of acidic or basic groups of the interfering
species. For example, tryptophan is often a fluorescing contaminant in biological samples. However, the intense fluorescence
of tryptophan (maxium at 348 nm) is quenched in solution at pH
below 1 (5a). This is attributed to protonation of both the
amino and carboxylic groups of the amino acid (6). Thus cin-
chonine, which fluoresces intense blue in acidic solutions, can
be determined unambiguously in the latter medium in presence of
tryptophan (7).

Furthermore, either on a trial and error basis or if the
identity of the fluorescing contaminant is known, chemical
modification of the analytical sample may be employed to selectively convert the fluorescent contaminant to a non-fluorescent
product. Such possibilities may obviate the necessity of physi-
co-chemical separation of the analyte from the contaminants.
For example, anthracene, which has a broad excitation spectrum
that ranges from 200 to about 370 nm, and a fairly broad fluorescence emission spectrum ranging from 365 to 490 nm, is a
troublesome molecule in the fluorimetric analysis of aromatic
hydrocarbon mixtures. In particular, the lowest energy excitation band (290 to 370 nm) of anthracene occurs in a region where
a number of aromatic hydrocarbons might be selectively excited
if anthracene was not present. Schenk and Wirz (8) have developed a method for fluorimetric determination of aromatic hydrocarbons by selective excitation in presence of anthracene. The
method involves removal of the low energy excitation band of
anthracene by the Diels-Alder reaction of this compound with
maleic anhydride to form a non-fluorescent adduct. This method
permits the fluorimetric measurement of pyrene, diphenylstilbene,
and perylene in the presence of the adduct.

For organic compounds which are non-fluorescent or weakly
fluorescent indirect fluorimetric methods are used. This
involves either the conversion of the analyte to a fluorescent
derivative using an appropriate reaction scheme or the utiliza-
tion of the capability of the analyte to influence (enhance or
quench) the fluorescence of a fluorescent dye. The indirect
method has the general advantage over the direct method in that,
provided that a suitable reactant can be found to produce a
fluorescent derivative, virtually any compound having functional
groups can be made amenable to fluorimetric analysis. However,
it should be borne in mind that derivatization methods are most
often applicable to classes of analytes rather than specific
analytes and that the fluorescence of the derivative usually
arises from an electronic transition localized on the derivatiz-
ing reagent. As a result, the gain in generality of the indirect

method is partially offset by the sacrifice of the analytical
selectivity.   Moreover, because the derivatization reaction is
at least bimolecular in the great majority of the cases, the
maximum obtainable fluorescence intensity from the derivative
will be limited by the concentration of the analyte, insofar
as it affects the completeness of the derivatization reaction.
The latter will generally decrease as the analyte concentration
decrease.   Consequently, indirect fluorimetry will in many cases
be less sensitive than direct fluorimetry unless the quantum
yield of the derivatized fluorophore is unusually large.

There are several books and reviews available on the application of luminescence spectroscopy for the qualitative and
quantitative analysis of organic compounds.   Fluorescence assay
methods of compounds of biological and medicinal interest are
discussed by Udenfriend in his two books (5a,5b).   Information
on luminescence characteristics of various classes of organic
compounds is presented in a book by Pringsheim (9).   Hercules
(10) has edited a book on principles and application of fluorescence and phosphorescence analysis.   White and Weissler (11,
12) in their two books provide reference material on the analysis
of organic compounds.   They have also published biannual reviews
(13,14) on fluorimetric analysis.   Berlman (15) has published a
handbook in which are presented fluorescence spectra of about
a hundred aromatic compounds.   Phillips and Elevitch (16) have
discussed fluorescence techniques of analysis in clinical pathology.   Passwater (17) has published a guide to fluorescence
literature.   The application of fluorescence techniques for the
assay of organic compounds is also discussed by Konstantinova-
Shlesinger (18).

Recently, Bartos and Pesez (19) have discussed luminescence
methods of qualitative and quantitative analysis of organic
compounds.   Winefordner and co-workers (20) have written a
chapter on the use of phosphorimetry for chemical analysis, in
which they have discussed in detail some of the applications of
phosphorimetry.   White and Argauer (21) in their book on

fluorescence analysis have discussed fluorimetric analysis of
selected organic compounds. Winefordner, Schulman and O'Haver
(22) have recently published a book on luminescence spectrometry
where they have discussed analytical use of fluorimetry and
phosphorimetry. Smith (23) has reported phosphorimetric limits
of detection of several biologically important compounds includ-
ing local anesthetics, metabolites and carcinogens. Guilbault
has edited a book (24a) and has authored another (24b), both on
the subject of fluorescence (theory, instrumentation and practical
application). The two books have details of the application of
fluorescence and phosphorescence spectroscopy.

Since the literature of the application of luminescence
spectroscopy is extensive and since several of the books of the
subject describe the details of luminescence analysis of organic
compounds, the present discussion will be limited to the few
examples illustrating the use of different techniques involved
in the development of a fluorimetric or a phosphorimetric method
of analysis for organic compounds. Fluorimetric and phosphorimetric analytical data for some organic compounds are listed in
Tables 4.1 and 4.2, respectively.

Methods Using Native Luminescence. The most desirable
method of luminescence analysis is the one that employs the
native fluorescence or phosphorescence of the analyte, as this
method is usually more sensitive than the derivatization method
and is certainly less complicated. An example of a fluorimetric
method of analysis based on the native fluorescence of the analyte
is the assay of lysergic acid diethylamide (LSD). LSD is highly
fluorescent with excitation maximum at 325 nm and emission maximum at 445 nm (25). Axelrod and co-workers (26) have devised a
method for the analysis of LSD in tissues which can determine as
little as 3 ng of the compound per gram of tissue. Agajanian and
Bing (27) applied the method of Axelrod (26) to analyze LSD in
plasma. The procedure is as follows. The homogenized tissue,
serum or urine (up to 5 ml) is mixed with 25 ml of $\underline{n}$-heptane
(containing 2 percent isoamyl alcohol) and 0.5 ml of $1\underline{N}$ NaOH.

The mixture is saturated with sodium chloride and shaken for 15 min. An aliquot of the heptane layer is withdrawn and shaken with 3 ml of 4×10^{-3} N HCl for 10 min. The heptane is then removed by aspiration and the fluorescence intensity of the acid phase is determined and compared with the standards analyzed in the same way. Metabolites of LSD do not interfere in the estimation.

Another example representative of direct fluorimetric analysis is the assay of quinine. Quinine is intensely fluorescent in protonated form (pH $\leq$ 1) with excitation maximum at 350 nm and emission maximum at 450 nm (5a). Udenfriend and co-workers (5b) have developed a fluorimetric assay for quinine in biological samples which is as follows. The homogenized tissue, serum or urine (up to 10 ml) is mixed with equal volume of 0.1 N sodium hydroxide and 30 ml of benzene and shaken for 30 min. The mixture is centrifuged and 1 ml of isoamyl alcohol is carefully added to the benzene layer. Twenty milliliters of the benzene layer is shaken with 3 ml of 0.1 N sulfuric acid for 3 min., centrifuged and the organic layer aspirated. The fluorescence intensity of the acid phase is then measured and compared with the standards carried through the same procedure. Metabolites of quinine do not interfere in this procedure and as little as 2 ppb can be estimated.

An example representative of the direct phosphorimetric method of analysis is the determination of aspirin in blood serum or plasma. Unlike salicylic acid which is strongly fluorescent and non-phosphorescent, aspirin is weakly fluorescent and strongly phosphorescent. This makes it possible to determine aspirin phosphorimetrically in the presence of salicylic acid and is an excellent example of the complimentary nature of fluorimetry and phosphorimetry. The phosphorimetric assay for aspirin, developed by Winefordner and Latz (28), is as follows. 0.4 ml of serum or plasma is placed in a 10 ml glass-stoppered graduated cylinder. Exactly 0.1 ml of concentrated hydrochloric acid and 7.5 ml of spectral grade chloroform

are then added to the blood sample in the cylinder. The
cylinder is shaken vigorously for 30 seconds and the solvent
layers are allowed to separate. A 1.0 ml aliquot of the organic
layer is pipetted into a 10 ml beaker and the chloroform is
rapidly evaporated by using a gentle stream of air or nitrogen
gas to flow over the surface of the solvent. The dry residue
thus obtained is then dissolved in 1.0 ml of EPA (5:5:2, volume
ratio of ether, isopentane and ethanol). The EPA solution is
then poured into a 3 ml clean, dry hypodermic syringe, and the
plunger replaced. Half of this solution is used for rinsing
the sample cell, and the other half is used for the measurement.
The sample cell is mounted in the Dewar flask containing liquid
nitrogen and aligned. The relative phosphorescence intensity
is determined and corrected for normal blood background. The
optimum concentration range for the phosphorimetric aspirin
analysis is 50-500 µg per ml of blood serum of plasma.

Another example of direct phosphorimetric analysis is the
assay of p-nitrophenol, (the major metabolic product of para-
thion) in human urine, developed by Moye and Winefordner (29).
This assay also demonstrates the use of thin layer chromatography (TLC) for the separation of an analyte prior to its
determination. The procedure is as follows. Ninety ml of urine
is acid hydrolyzed with 10 ml of concentrated HCl. 5.0 ml of
hydrolyzed urine is mixed with 6 ml of ether, shaken vigorously
for 5 minutes, and the aqueous layer is removed and discarded.
The ether layer is evaporated to 2.0 ml and then applied to a
TLC plate coated with activated silica gel G. After developing
the plate by the ascending method using ether, the spot corresponding to p-nitrophenol appears about halfway up the plate and
directly above a strong blue fluorescence band which shows up
under ultraviolet light. All of the thin layer on the TLC
plate except the spot corresponding to p-nitrophenol is scraped
off and discarded. The thin layer containing the p-nitrophenol
spot is then scraped into a capped vial, 5 ml of 0.1 M HCl is
added, and the contents of the vial shaken vigorously for 10
minutes. The slurry in the vial is centrifuged, and the clear.

supernatant liquid is removed and extracted with several small
portions of ether. The ether extracts are combined and diluted
to 10.0 ml with ether. The phosphorescence of the ether solution
is determined using excitation wavelength of 265 nm and emission
wavelength of 525 nm. The measured phosphorescence is corrected
for the blank reading which is obtained by carrying a urine
sample known to contain no drugs, through the same
procedure. The concentration of p-nitrophenol is determined by
using an analytical curve prepared with standard solutions of
the compound. The entire procedure requires 40 minutes and the
minimum concentration of p-nitrophenol that can be analyzed by
this method is 0.01 µg in a 5 ml sample of urine.

Indirect Methods

Derivatization. Conversion of a non-fluorescent
analyte to a fluorescent derivative for its fluorimetric determination has been achieved by several techniques. Oxidation,
for example, is used to convert phenothiazines into corresponding
sulfoxides which are highly fluorescent (30-32). Mellinger and
Keeler (30,31) who studied all aspects of the fluorescence of
thorazines and their oxidized products, reported potassium
permanganate to be unique for this type of reaction. However,
Ragland and Kinross-Wright (32) found that hydrogen peroxide
in 50% acetic acid was also a good oxidant for the conversion
of phenothiazines to the fluorescent sulfoxides.

Ragland and Kinross-Wright (33) have developed a procedure
for determining phenothiazines and their sulfoxide metabolites
in plasma or tissue at levels as low as 0.05 µg per ml for
thioridazine and 1 µg/ml for chlorpromazine. One ml of plasma
or tissue homogenate is mixed with 1.5 ml of 3 N NaOH in a 12
ml glass-stoppered centrifuge tube. To this mixture (in the
case of chlorpromazine assay the mixture must be heated at 100°
C for 5 minutes and then cooled) 6 ml of n-heptane containing
1.7% isoamyl alcohol is added, the tube is shaken for 10 minutes
on an automatic shaker, and then centrifuged. Five milliliters
of the organic layer is transferred to another tube and shaken

with 2 ml of pH 5 buffer to extract sulfoxide metabolites. A
4.0 ml aliquot of the heptane layer is transferred to another
tube and shaken with 2 ml of 50% acetic acid to extract the
parent phenothiazine. The residual heptane in the buffer extract of metabolites and in the acetic acid extract of the phenothiazine is removed by aspiration. To the buffer extract equal
volume of glacial acetic acid is added. To both extracts one
tenth volume of 3% hydrogen peroxide is added. The samples are
heated for 10 minutes in a boiling water bath and cooled quickly
to room temperature. The fluorescence intensities of the two
samples are measured with λ_{ex} = 340, λ_{em} = 380 for chlorpromazine and λ_{em} = 440 for thioridazine. Blanks and standards are
carried through the entire extraction procedure.

Cyanogen bromide is another oxidizing agent commonly used
for oxidizing non-fluorescent or weakly fluorescent analytes to
analytically useful fluorophores. Chlorothen, methylpyrilene,
pyrilamine, thenyldiamine, and tripelennamine all yield fluorescent products on treatment with cyanogen bromide in neutral
solution for 30 minutes at room temperature (34). The products
all show maximal excitation at 345-355 nm and maximal fluorescence at 412-419 nm. Isoniazid reacts with cyanogen bromide
under specific conditions to give an intensely fluorescent
product. Peters (35) used this reaction coupled with solvent
extraction to develop a specific assay procedure for the determination of isoniazid in plasma with no interference from
nicotinamide.

Some compounds can be converted to fluorescent derivatives
by treating them with strong mineral acids. For example,
morphine gives a highly fluorescent derivative after heating it
in concentrated sulfuric acid. The reaction product fluoresces
(λ_{ex} = 290, λ_{em} = 400 nm) when the solution is made alkaline
(36). Codeine and codethyline when heated in concentrated
sulfuric acid for 3 minutes at 100°C and then made alkaline
show analytically useful fluorescence at 478 nm and 499 nm,
respectively (37).

Cholesterol when reacted with a mixture of acetic anhydride, trichloromethane and concentrated sulfuric acid yield a strongly fluorescent fluorophore (38). McDougal and Farmer (39) used this reaction to develop a fluorimetric assay for cholesterol requiring only micro-amounts of serum. The procedure is as follows. To 20 µl sample of serum, placed in a 15 ml centrifuge tube, is added 0.2 ml of acetic acid-chloroform (3:2). A momentary precipitate forms which is redissolved by mixing. Then 5 ml of the freshly prepared chloroform-acetic anhydride (10:3) reagent is added and mixed. To this solution 0.2 ml of concentrated sulfuric acid is added. The solution is immediately mixed, centrifuged, and the clear supernatant is decanted into the fluorescence cell. The fluorescence intensity is measured (λ_{ex} = 546, λ_{em} = 600 nm) 40 minutes after the addition of sulfuric acid. For the blank, 20 µl of distilled water is used. Standard solutions and the blank are carried through the same procedure as the serum sample. The concentration of cholesterol is determined by comparing the fluorescence intensity of the sample, corrected for blank reading, with the intensity of the standard solutions. This procedure has been widely adopted by clinical laboratories.

Weakly fluorescent or nonfluorescent primary arylamines and arylhydrazines can be condensed with aldehydes or ketones to often yield fluorescent derivatives. A general procedure for assaying primary amines using fluorescamine (4-phenyl-spiro(furan-2(3H),1'-phthalan)3,3'dione) reagent has been developed by Udenfriend and coworkers (40,41). Fluorescamine is a highly sensitive fluorogenic reagent which reacts directly with primary aliphatic or aromatic amines to form a fluorophore of high intrinsic fluorescence. Fluorescamine has found extensive application in a wide variety of biochemical and biophysical studies (42-47). It has also been used as a fluorimetric spray reagent for thin layer chromatography (48), and in clinical toxicology in the fluorimetric detection of drugs of abuse such as amphetamine, in urine (49). Recently de Silva and Strojny (50) have published results of a systematic

examination of a number of pharmaceutical compounds having either
a primary aromatic or aliphatic amine group which were reacted
with fluorescamine in solution and on thin layer chromatographic
plates. They have presented the analytical parameters for opti-
mal reaction and sensitivity which can be utilized for the
determination of compounds studied in biological fluids.

   Ninhydrin (1,2,3-indantrione) is another sensitive fluorimetric reagent that reacts with primary amines to form highly
fluorescent condensation products. It has been widely used for
the determination of amino acids, amines, amino sugars, and
peptides with nanomole sensitivity (51). It has also been used
as a spray reagent for thin layer chromatographic analysis.
The general procedure for the determination of amino acids with
ninhydrin is as follows (52). To one millilitre of a solution
of amino acid (containing 0.1 to 10 μg of amino acid and pre-
adjusted to pH 8.5), are added 0.5 ml of 0.2 M n-butyraldehyde
solution in water and 0.5 ml of 14×10^{-3} M (in 0.1 M carbonate
buffer, pH 8.5) ninhydrin solution. The mixture is incubated
at 60°C for 45 minutes, then cooled and diluted with an equal
volume of doubly distilled water. All amino acids and amines
which react show an excitation maximum at 385 nm. The fluorescence emission maximum varies from 450 to 495 nm, depending
on the analyte. The relative intensities of fluorescence
obtained for different amines and amino acids vary markedly
with the reaction conditions. Thus, using the reaction conditions as originally worked out by Lowe and co-workers (52),
γ-aminobutyric acid is 20-100 times as fluorescent as most α-
amino acids, and the procedure can be used to assay for γ-amino-
butyric acid in tissue. McCaman and Robins (53) modified the
ninhydrin reaction conditions so as to obtain the most intense
fluorophore with phenylalanine. Wong and co-workers (54)
modified McCaman and Robins (53) ninhydrin reaction and used it
for developing a highly specific and sensitive fluorimetric
method of analysis for phenylalanine in 25 μl samples of blood.

   Aldehydes and ketones of analytical interest which do not

fluoresce, may be condensed with arylhydrazines, such as naphthylhydrazines, or with arylamines to yield fluorescent products. Thus acetaldehyde can be analyzed fluorimetrically by condensing it with 3,5-diaminobenzoic acid (55). Camber (56) has proposed salicyloyl hydrazide as a reagent for carbonyl compounds. Brandt and Cheronis (57) used the reagent 2-diphenyl-acetyl-1,3-indandione-1-monohydrazone for the assay of carbonyl compounds. Sawicki and co-workers (58,59) have published surveys of reagents for the fluorimetric assay of formaldehyde. Nanogram quantities of formaldehyde can be detected by condensation with 2-nitro-1,3-indandione. Sawicki and co-workers (60) have also reported several fluorimetric methods for the assay of β-hydroxyacrolein -- a tautomer of malonandehyde. This compound reacts with many amines to give intense fluorophores, of which the products formed with p-aminobenzoate (λ_{ex} = 500, λ_{em} = 550 nm) and 4,4'-sulfonyl-dianiline (λ_{ex} = 475, λ_{em} = 545 nm) are the most intense and stable.

α-Keto acids can be condensed with o-phenylenediamine to yield stable and highly fluorescent quinoxaline derivatives. Pyruvic acid, for example, reacts quantitatively with o-phenylenediamine to give the fluorophore, 2-hydroxy-3-methylquinoxaline (61). Spikner and Towne (61) have developed the following assay procedure for keto acids. Aliquots of a keto acid sample (containing 0.05 - 2.0 μg) in 2 N H_2SO_4 are transferred to tubes containing 1 ml of freshly prepared o-phenylenediamine solution (10 mg of the compound in 2 N H_2SO_4) and the volumes adjusted to 3 ml with 2 N H_2SO_4. The tubes are heated on a boiling water bath for 2 hours, cooled, and 1.66 ml of cold concentrated sulfuric acid are added. The cooled samples are adjusted to 5.0 ml with 50% H_2SO_4 and their fluorescence intensities measured and compared with the standards, carried through the same procedure, to determine the concentration of the acid. The wavelengths of excitation and emission depend on the type of α-keto acid to be analyzed, and fall in the range of 360-368 nm and 490-503, respectively, for eight of the nine acids studied by Spikner and Towne (61). Glyoxylic acid,

however, had the excitation maximum at 338 and emission maximum
at 518 and gives the least intense fluorophore of the nine acids.

Hydrolysis of easily hydrolyzed non-fluorescent compounds
may produce, in certain cases, an analytically useful fluorophore.
For example, acetylsalicylic acid (aspirin) has weak native
fluorescence, but its base-hydrolysis product, salicylate ion,
is strongly fluorescent (λ_{ex} = 313, λ_{em} = 442 nm). This fact
is used in developing fluorimetric methods of analysis for
aspirin in serum (62,63) and urine (64). Several pesticides
(Benomyl, Co-Ral, Coumaphos, Guthion, Maretin, Zinophos) undergo
base-hydrolysis to yield highly fluorescent products which can
be used for the fluorimetric analysis of the parent pesticide.
For example, Co-Ral is hydrolyzed by hot alkali to yield highly
fluorescent coumaric acid or coumaritic acid (65,66). Mac-
Dougall and co-workers have developed a fluorimetric method of
analysis of Co-Ral at levels as low as 0.02 ppm in a 50 g sample
of animal tissue (67). The procedure is given below.

A 50 g sample of animal tissue is extracted with 200 ml of
acetone and 5 g of Super-Cel in a blender, for 5 minutes, and
then filtered. The residue is reextracted with 200 ml of benzene
in a blender, for 5 minutes, and filtered. The acetone and
benzene filtrates are combined in a separatory funnel and shaken
vigorously. The lower aqueous phase is discarded. To the ben-
zene-acetone solution, 5 g of Super-Cel and 200 ml of chloroform
is added. The slurry is mixed thoroughly, and filtered. The
solvent is evaporated on a steam bath, the resulting residue
dissolved in 300 ml of Skellysolve B, and the solution extracted
three times with 50 ml portions of acetonitrile. Each acetonitrile fraction is washed with a 200 ml portion of Skellysolve B.
The three fractions are then combined and the solvent evaporated
to dryness. The residue obtained is dissolved in 10 ml of
chloroform. Co-Ral is further separated and concentrated by
using column chromatography as follows. A column is prepared
by packing a 20 x 400 mm chromatographic tube with 5 g of
Super-Cel. A slurry of 30 g of alumina (acid-washed) in chloroform is poured down the column and the tube washed with 100 ml

of chloroform. To the column 10 ml of a chloroform solution
of the residue obtained in the extraction step is added and Co-
Ral is eluted with 100 ml of chloroform. The eluate is then
concentrated to about 10 ml, and then diluted to exactly 25 ml.
A 5 ml aliquot of this solution is concentrated in a screw-cap
tube to near dryness on a steam bath. To the tube 5 ml of 1 N
KOH is added. The tube is capped and placed in an oven at $92^{\circ}C$
for 2.5 hr. The solution is then cooled and extracted with 5
ml of amyl alcohol and the alcohol layer discarded. The fluores-
cence intensity (λ_{ex} = 330, λ_{em} = 410 nm) of the aqueous solution
is measured and the concentration of Co-Ral is determined by
comparing the fluorescence intensity of the sample with that for
Co-Ral standards hydrolyzed in the same way.

Chelation of the organic analyte with certain metal ions
yields analytically useful fluorescent chelates. For example,
the magnesium chelates of tetracyclines are highly fluorescent
and some tetracycline-metal complexes have been used for the
fluorimetric analysis of the drugs (68,69). A highly sensitive
and specific method of analysis for the tetracycline antibio-
tics (with the exception of oxytetracycline) in body fluids and
tissues was reported by Kohn (70), based on the ability of
tetracyclines to form highly fluorescent, extractable mixed
complexes with calcium and barbiturates. The general procedure
is as follows. An aliquot of sample (body fluid or tissue extract in 0.1 N HCl) is diluted with water to 4.00 ml, and 1.00
ml of a solution of 9.1 N (1.5 N in the case of tissue extract)
trichloroacetic acid and 0.004 M KH_2PO_4. The mixture is centrifuged to precipitate proteins and 4.00 ml of supernatant are
mixed with 2.00 ml of a solution of 0.06 M $Pb(NO_3)_2$ and 2.0 M
sodium acetate. Four millilitres of supernatant are mixed with
1.00 ml of 0.32 M KIO_3 and centrifuged after 60 minutes. Four
millilitres of the supernatant are added to a glass-stoppered
centrifuge tube containing 3.00 ml of ethyl acetate, 3.0 ml of
0.9 M sodium barbital, and 3.0 ml of 0.1 M $CaCl_2$. The tube is
shaken vigorously for about 2 minutes and the phases are allowed
to separate. The fluorescence intensity (λ_{ex} = 405, λ_{em} = 530

nm) of the upper layer is measured and compared with standards
to determine the amount of tetracycline in the sample. As
little as 0.1 M μmole (0.05 μg) of tetracycline could be determined with a standard deviation of about 10%.

Some organic compounds that have basic groups can interact
with water soluble acidic fluorescent dyes, such as eosin, to
form fluorescent neutral complexes which will be extractable
into organic solvents. Since the dyes are anions, the organic
solvent extract will contain only the neutral complex, and
therefore, the fluorescence intensity of the extract will be a
measure of the analyte. Glazko and co-workers (71) have investigated a variety of acid dyes and conditions for their application to the assay of basic drugs. They were able to develop an
assay for plasma levels of an antihistamine, diphenylhydramine
(Benadryl) by using the fluorescent dye Tinopal GS. The procedure is as follows.

To a sample of plasma (1-10 ml, containing 0.2-2 μg of
diphenhydramine), an equal volume of 0.1 N NaOH and 25 ml of
purified heptane is added, the mixture shaken and centrifuged.
To the heptane layer 0.2 ml of ethanol is added to prevent
adsorption of the analyte onto the glass surfaces of the tube
and pipette. Twenty ml of heptane layer are then transferred
to another tube containing 3.5 ml of 0.1 N HCl, the tube shaken,
centrifuged and the heptane layer discarded. Three ml of the
acid layer is transferred to another tube containing 0.2 ml of
0.5 N NaOH in half-saturated sodium carbonate and 1.3 ml of 1,2-
dichloroethane, the tube shaken, centrifuged and the aqueous
layer removed by aspiration. The residual organic layer is
poured into another tube (taking care not to transfer any drop-
lets of alkali), one drop of 10% Tinopal GS in 0.1 M citric
acid is added and the tube is shaken. The amount of dye extracted into the 1,2-dichloroethane layer is proportional to
the amount of the drug. The tube is centrifuged and the
fluorescence intensity (λ_{ex} = 365 and λ_{em} = 450 nm) of the
organic layer (to which 0.2 ml of methanol should be added to

prevent adsorption of dye complex onto the glass surfaces) is measured and compared with standards and blanks carried through the same procedure. As little as 0.08 µg of diphenhydramine per ml can be determined by this procedure.

Generally, an organic base-dye complex will have spectral characteristics different from that of the dye itself, in addition to having a lower solubility in polar solvents. This spectral change can be utilized for the fluorimetric assay of the complex. Laugel (72) has utilized such a procedure for assay of basic drugs. Ogawa and co-workers (73) have used the procedure of Laugel (72) to assay atropine by complexing it with the acidic dye, eosin Y (tetrabromofluorescein) and measuring the fluorescence intensity (λ_{ex} = 365, λ_{em} = 556 nm) of the complex in chloroform. As little as 1 µg/ml of atropine can be assayed.

Cohen (74,75) used the dye-complexing procedure to develop a fluorescence assay for d-tubocurarine. In his first report (74) he described a general procedure for the assay of d-tubocurarine using the dye rose bengal. In a subsequent paper (75) he described individual extraction procedures to determine d-tubocurarine in body fluids (cerebrospinal fluid, urine and plasma) and tissues (liver, kidney, muscle, brain and fat). Sensitivity of the assay varied with the kind of biological sample. In urine, the limit of detection was 0.05 µg/ml of the drug.

Certain organic compounds may interact with intensely fluorescent dyes and quench the fluorescence of the latter. The extent of quenching can be quantitatively related to the concentration of the quencher. For example, Sturgeon and Schulman (76) have developed a quenchofluorimetric method of analysis for arylamines using the intensely fluorescent Bratton-Marshall reagent (N-1-naphthylethylenediamine). The procedure involves diazotization of a weakly or non-fluorescent arylamine and then condensing the product with Bratton-Marshall reagent. This produces a non-fluorescent diazo derivative, thereby

quenching the fluorescence of the reagent. Thus, by quantitat-
ing the disappearance of the fluorescence of Bratton-Marshall
reagent the concentration of the arylamine can be determined.

Karust and co-workers (77) noted that the fluorescence of
fluorescein mercuric acetate (λ_{ex} = 499, λ_{em} = 520 nm) in 1 N
NaOH is quantitatively quenched by disulfide groups. They
applied this fluorescence quenching to develop an assay proce--
dure to determine the number of disulfide groups in proteins
and peptides.

Enzymes have been used as reagents for indirect fluorimetric
analysis. Because enzymes are biological catalysts capable of
catalyzing a specific reaction of a particular substrate, they
are of great use in chemical analysis.

Guilbault (78) in his book, Enzymatic Methods of Analysis,
has discussed the uses of enzymes in the analysis of substrates,
activators, and inhibitors. Fluorimetric methods of analysis
of enzymes and substrates are discussed by Udenfriend (5b),
Purdy (79), and Guilbault (23,24). Phillips and Elevitch (16)
have discussed fluorimetric technqiues in clinical pathology
in the book, Progress in Clinical Pathology, edited by Steffani.

The principle underlying the enzymatic analysis is the
Michaelis-Menten equation for the enzyme kinetics:

$$E + S \underset{k_{-2}}{\overset{k_1}{\rightleftharpoons}} ES \underset{k_{-2}}{\overset{k_2}{\rightleftharpoons}} E + P$$

where E is an enzyme that combines with a substrate S to form
an intermediate complex ES which subsequently breaks down into
products P, liberating the enzyme. The equilibrium constant,
K_{eq} for the formation of the complex (also known as the
Michaelis constant) is defined as,

$$K_{eq} = \frac{k_2 + k_{-1}}{k_1}$$

The rate of the reaction is a function of the concentrations of enzyme and substrate. In presence of activator (A_c) and inhibitor (I_n) the rate is also a function of A_c and I_n concentrations. For a given enzyme concentration, the initial rate of the reaction, r_o, is given by the relation:

$$r_o = \frac{r_{max}[S]_i}{K_{eq} + [S]_i}$$

where r_{max} is the maximum reaction rate and S_i is the initial substrate concentration. As S_i is increased, the initial rate increases until a nonlimiting excess of substrate is reached (i.e. $S_o \gg K_{eq}$), after which additional substrate causes no increase in the reaction rate.

The concentration of substrate in an enzymatic reaction can be determined either by measuring the total concentration of the product formed (equilibrium method) or by measuring the rate of the enzyme reaction (kinetic method). In the first method, a large excess of enzyme over the substrate concentration is used to ensure a relatively complete reaction. After the reaction reaches equilibrium, the amount of product formed is measured. From this, the substrate concentration can be determined. For example, glucose can be selectively determined in a mixture of carbohydrates by using an enzyme reaction with glucose oxidase to produce hydrogen peroxide (80). The amount of H_2O_2 produced is measured fluorimetrically by using its reaction with the enzyme peroxidase and p-hydroxyphenylacetic acid to produce a fluorophore (λ_{ex} = 317, λ_{em} = 414 nm).

$$\text{glucose} + H_2O + O_2 \xrightarrow{\text{glucose oxidase}} \text{gluconic acid} + H_2O_2$$

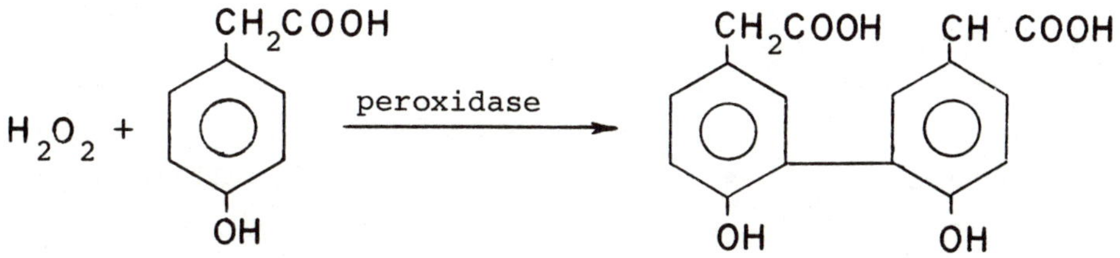

H₂O₂ + [CH₂COOH / OH phenol ring] →(peroxidase)→ [CH₂COOH / OH ring]—[CH COOH / OH ring]

(nonfluorescent)   (fluorescent)

In the kinetic method, the initial rate of reaction, r_o, is measured in the usual manner, i.e. by following either the rate of formation of product or the rate of disappearance of the substrate. Since r_o is a function of the concentration of the substrate for fixed concentrations of enzyme, activator and inhibitor, the concentration of the substrate can be determined directly from r_o. For example, the concentration of glucose in the above reaction can also be determined by measuring the initial rate of formation of the fluorophore.

The kinetic method of enzymatic analysis, depending on reaction conditions, can be used for the determination of substrate, enzyme, activator or inhibitor concentration, while the equilibrium method is useful for the determination of substrate only. Moreover, the kinetic method is faster, because the rate can be measured initially without having to wait for the reaction to go to completion. Care must be taken, however, to control the pH, temperature and ionic strength of reaction medium to obtain maximum sensitivity in the kinetic method. Guilbault and co-workers (81,82) and Pardue and co-workers (83) have been able to obtain, with reasonable care, precision and accuracy of better than 1% in the kinetic method of enzymatic analysis.

Inorganic Substances

Fluorescence spectroscopy is used extensively in the qualitative and quantitative analysis of inorganics. The application
of phosphorescence to inorganic analysis is not common, but,
undoubtedly, with the advent of new techniques and instrumentation phosphorimetry will become as useful a tool as fluorimetry
for inorganic analysis. Several inorganic substances are fluorescent or phosphorescent in the solid state, but only a few
can be analyzed fluorimetrically or phosphorimetrically directly
in the solid state. Inorganic substances that can be analyzed
by fluorimetry in solution are all either metal ions or inorganic
anions. There are, as for organic compounds, two methods of
fluorimetric analysis of inorganics, direct and indirect. In
the direct method the native fluorescence of the analyte is used
for quantitation. In the indirect method, a non-fluorescent or
a weakly fluorescent analyte is converted, with a suitable
reagent, to a fluorophore and the fluorescence intensity of the
fluorophore is quantitated, or the extent of quenching of an
intensely fluorescent compound by the analyte is used for quantitation.

The application of fluorescence spectroscopy to the
analytical problems in inorganic chemistry has been discussed
in several books (5a,5b,10,18,21,24a,24b,163). Weissler and
White have written biannual reviews on fluorimetry (13,14) in
which fluorimetric analysis of inorganic substances have been
discussed. They have also written a chapter on the subject in
the book, Handbook of Analytical Chemistry, edited by Meites
(11).

Direct Methods. There are only a few inorganic compounds and ions that possess native fluorescence. Therefore,
the direct fluorimetric method is limited in its application
to inorganic analysis. Several inorganic compounds and mixed
salts are fluorescent or phosphorescent. These types of compounds fall into two fundamentally different classes; chemically
pure compounds and multicomponent systems. A survey of chemically

pure inorganic compounds that possess native fluorescence is
presented in a review by Randall (164). Examples of this class
of compounds are the salts of the rare earth elements, uranyl
salts, manganese halides, tungstates and molybdates. In addition to these there are certain salts which although they are
non-fluorescent at room temperatures, become fluorescent at
cryogenic temperatures. Examples of these salts are silver
iodide, mercuric iodide, lead iodide and cadmium sulfide (165).

The multicomponent inorganic compounds, which are also
known as phosphors, are polycrystalline substances with trace
amounts of foreign ions (impurities) which act as activators
for the observed luminescence. Examples of polycrystalline
materials that show phosphor properties are zinc, cadmium,
calcium and strontium sulfides, potassium chloride, zinc sele-
nide, calcium and magnesium tungstates, beryllium, zinc and
cadmium silicates and many others. Foreign ion activators
include copper, silver, manganese, antimony, thallium, lead,
bismuth, rare earths and uranium. The preparation and properties of inorganic phosphors are discussed by Pringsheim (9) and
by Leverenz (166) in their books. Because the luminescence of
phosphors is usually critically dependent on the precise conditions under which they are prepared, phosphors are used in
few analytical applications.

Inorganic compounds that show native fluorescence in
solution are restricted to the ions of a few elements, notably
uranium and the rare earths (9). Even among uranium compounds
only those with the uranyl ion (UO_2^{+2}) are fluorescent. The
fluorescent properties of a number of uranium compounds have
been studied by Vavilov and Levshin (167). Uranium fluoresces
in sodium fluoride beads and this is used to develop sensitive
fluorimetric methods for determination of uranium in various
kinds of samples (168,169,170). For example, Steele and Robert
(171) determined small quantities of the uranyl ion by extract-
ing uranyl nitrate into ethyl acetate in the presence of
aluminum nitrate. An aliquot of the organic layer is then

pipetted into a pellet of sodium fluoride contained in a
platinum dish and is fused after evaporation. The fluorescence
of the fused cake is measured. As little as 30 ppb of uranium
can be determined by this method. White (172,173,174) has
suggested the use of the fluorescence of uranium in concentrated
acids for its determination. For the microdetermination of
uranium in biological material, the uranium should be first
precipitated as an uranium-protein complex (172).

Pringsheim (9) and Zaidel and Larionov (175) have discussed
the fluorescence emission spectra of solutions of inorganic
salts of the rare earths. Because the fluorescence of the rare
earth ions involves the transition of $\underline{f}$ electrons which are well
shielded from external influences, the spectra are usually sharp
(similar to atomic spectra). Terbium, gadolinium and cerium
display the strongest fluorescences in solution among the rare
earth ions (175). The fluorescence spectra of cerium salts
consist of a broad band extending from 330 to 402 nm. As little
as 10 ppb of cerium can be determined fluorimetrically. Terbium
salts fluoresce yellow-green and concentrations of 1 ppm to 10
ppb can be determined. Europium fluoresces red and its minimum
detectable concentration is 10 ppm. Cerium (III) exhibits
characteristic fluorescence (λ_{ex} = 258, λ_{em} = 350 nm) in dilute
inorganic acid solutions. M. Furukawa and co-workers (176)
and Poluektov and co-workers (177) have used this property of
Ce (III) to develop a fluorimetric assay procedure for Ce (III)
in the presence of lanthanum, yttrium, and europium. NO_3^-,
Fe^{3+} and Ce^{4+} are the potential interferences (177). As little
as 1 ppm of Ce (III) can be determined by this method. Fassel
and Heidel (178,179) have developed sensitive fluorimetric
methods of analysis for the rare earth ions. For example,
terbium can be determined in hydrochloric acid solution (using
hydrogen arc lamp for excitation) down to a concentration of
about 4 µg of terbium per ml.

The qualitative analysis of the rare earth elements can
be conveniently performed by observing the color and number of
fluorescence bands of rare earth elements in borax and phosphoric

acid beads (180). The detection limits for the fluorescence
analysis of rare earth elements by this method, as reported by
Haitinger (180), are: cerium, 0.4 µg; samarium, 4.5; europium,
20.0; gadolinium, 45.0; terbium, 2.0; and dysprosium, 4.5.
Neunhoeffer (181) has described a basically different method
for the general detection of rare earth elements. The method
involves introducing calcium oxide containing the sample into
the lower part of a colorless hydrogen flame. The presence of
a rare earth element is indicated by a flare-up of fluorescence.

Few direct fluorimetric methods for other inorganic compounds have been described. Belzi and Kishnirenko (182) have
reported fluorimetric assay procedures for As (III) and As (V)
in hydrochloric and hydroleromic acid, respectively. The fluorescence is measured in frozen solutions at -196°C. In 7.6 M
HCl as little as 0.15 ppm of As (III) and 37 ppm of As (V) could
be determined. The reported limit of detection for the two ions
in 7.6 M HBr is 7.5 ppb for As (III) and 3.7 ppm for As (V).
Belzi and Kishnirenko (183) have also worked out similar assay
procedures for bismuth, antimony (III) and selenium (IV). In
hydrochloric acid solutions the reported limits of detection
are 2.1 ppb for bismuth, 1 ppb for antimony and 60 ppb for
selenium.

Bismuth, lead and antimony can be assayed fluorimetrically
in concentrated hydrochloric acid at -196°C (184). Kirkbright
and Saw (185) have developed the fluorimetric method of analysis
for lead. The fluorescence of lead is measured in hydrochloric
acid (3:10) containing 0.8 N potassium chloride within 15 minutes of mixing, with excitation wavelength of 270 nm and
emission wavelength of 480 nm. The range of concentrations
that could be determined by this method is 0.1 to 0.6 ppm of
lead. Bi, Cr^{6+}, Cu^{2+}, Fe^{3+}, Mo^{6+}, Tl^{+}, V^{5+}, ascorbic acid and
$S_2O_5^{-}$ were reported to be the potential interferences in the
determination of lead. Belzi and Kushnirenko (186) studied
the fluorescence of lead in "glasslike" frozen solutions of 8
N hydrochloric acid, hydrobromic acid, lithium chloride and

lithium bromide at -196°C. In 8 N HCl lead fluoresces at 423
nm when excited at 272 nm, and this fluorescence can be used to
determine as little as 10 ppb of lead. Hydrobromic acid or
lithium chloride can also be used as solvent for the determination of lead.

Tellurium (VI) exhibits fluorescence (λ_{ex} = 380, λ_{em} =
586 nm) at -196°C in 9 M hydrochloric acid "glass". Based on
this, Kirkbright and co-workers (187) developed a fluorimetric
method for the determination of tellurium in lead. This method
was sensitive down to 0.02 ppm of tellurium.

Thallium (I) and thallium (III) both can be assayed fluorimetrically at liquid nitrogen temperature in 3 N hydrochloric
acid saturated with sodium chloride with a sensitivity of 0.01
ppm of Tl (I) or Tl (III) (188). Thallium (I) can also be
determined in 8 N hydrochloric acid (λ_{ex} = 242, λ_{em} = 396 nm),
hydrobromic acid (λ_{ex} = 223,262, λ_{em} = 428 nm) or lithium
chloride (186). As little as 50 ppb of Tl (I) can be determined
by this method.

Recently, Shcherbov, Astaf'eva and Plotnikova (189) have
published a review in the Russian language on the fluorimetric
determination of As, Sb, Pb, Tl, Bi, Se and Te at -196° in frozen solutions of 4 to 9 N hydrochloric acid or hydrobromic acid.
The necessary techniques for the assays are discussed. The
same workers have published data on optimum conditions and
excitation and emission wavelengths for cryogenic fluorimetry
of Te, Pb, and Bi in 4.5 N hydrochloric acid, with good linearity
and a sensitivity of 0.1 µg/ml (190).

The recent growth of upper atmosphere and space chemistry
has stimulated work in gas phase fluorimetric analysis. For
example, the far infrared fluorescence of sulfur dioxide (191)
has been made the basis of a SO_2 detector which uses excitation
by the Zn 215-nm line or the Cd 229-nm line, the latter giving
a linear relationship between fluorescence intensity and

concentration over the range 0-1 to 1600 ppm (192). The
resonance fluorescence of nitric oxide at 215 nm, when excited
by sunlight, has been used to measure the nitric oxide concentration at many points in the upper atmosphere from an orbiting
satellite (193).

Indirect Methods. Indirect fluorimetric methods for
the analysis of inorganic substances are analogous to the indirect methods for the determination of organic compounds, and
are used for the determination of non-fluorescent inorganic ions
in solution. This method involves reacting the analyte with an
appropriate reagent to yield either a highly fluorescent product
from a non-fluorescent or weakly fluorescent reagent, or a less
fluorescent product from a highly fluorescent reagent. In the
former case the enhancement of fluorescence intensity of the
reagent is quantitated while in the latter case the quenching of
the fluorescence of the reagent is related to the concentration
of the analyte. The reactions that are commonly used in the
indirect methods are: oxidation, reduction, catalysis and
chelation. Of these, chelation enjoys the widest application
in the fluorimetric analysis of metal ions. Over forty different
metal ions have been determined by this method.

The chelation reaction involves combination of a metal ion
with a non-fluorescent or a weakly fluorescent aromatic compound
containing chelate-forming, electron-donating functional groups,
to form a highly fluorescent metal chelate. There are several
aromatic compounds which form fluorescent chelates with metal
ions. Some of these ligands are very specific for certain metal
ions under given reaction conditions, and therefore, yield highly
selective fluorimetric methods of analysis for these ions. An
example of a compound that has been extensively used as a ligand
for the fluorimetric analysis of metal ions by chelation is 8-
hydroxyquinoline (oxine). It forms fluorescent chelates with
Al, Be, Ca, Cd, Cs, Ga, In, Li, K, Mg, Na, Rb, Y and Zn, and
methods have been reported for the analysis of these metals
with oxine.

Because, in general, fluorescent chelates are formed only
by diamagnetic metal ions and not by paramagnetic transition
metal ions, the chelation method of fluorimetric analysis is
limited in its application to the former class of compounds.
However, this limitation may be viewed as an advantage in terms
of avoiding interference by the chelates of transition metals
in the determination of other metals.

For the analysis of an inorganic analyte in a mixture,
separation of the analyte by extraction or chromatography may
be necessary, prior to the fluorimetric determination, so as to
avoid interference from the impurities. In some cases, however,
depending on the fluorimetric reagent, manipulation of the
reaction conditions, such as pH and solvent, and proper choice
of excitation and emission wavelengths may remove the effects
of interferences and obviate the necessity of the separation
step.

Table 4.3 lists some typical fluorimetric methods and reagents for sixty-two inorganic ions. Since the details of these
methods (separation techniques, if any, solvent systems, solutions parameters, excitation and emission wavelengths, possible
interferences, etc.) vary with the analyte, the reagent and the
nature of analytical sample, and since these details are critical
in obtaining the reported sensitivity, the reader is advised to
go to the respective references to obtain the details of the
listed methods. The listing in the Table is not meant to be
exhaustive, but rather illustrative of the wide application of
fluorescence spectroscopy in inorganic analysis.

Some Specialized Applications

Metal Ion Coordination. The complexation of metal ions,
by monodentate or polydentate ligands, can be studied fluorimetrically in much the same way that it has been approached
absorptiometrically for the past several decades. The study
of complex formation invariably entails the establishment of

the stoichiometry of the complex or complexes (i.e. the number
of ligands bound to each metal ion) and the stability of each
complex (i.e. the formation constants). The absorptiometric
study of complexation is based upon the distinctions between the
absorptivities of the free ligand and the various complexes at
the analytical wavelength of absorption during the course of
titration with the metal ion (407-410). These features enable,
with the aid of Beer's law and the mass-balance expressions for
ligand and metal ion, respectively, the calculation of the
fractions of free and bound ligands (and therefore, absolute
concentrations of free and bound ligands if the formal ligand
concentration is known) and the average number of ligands bound
($\bar{n}$) at any point in the titration (411). If the complexes of
the ligand with the metal ion are relatively strong (formation
constants of 10^4 or greater), the maximum number of ligands
bound to the metal ion can be evaluated from a plot of absorbance
vs. mole-ratio of ligand to metal ion (or metal ion to ligand)
in which the concentration of either ligand or metal ion is
varied and that of the other component kept constant. This is
known as the mole-ratio method (407). Alternatively, Job's
method (412) or its modification by Vosburgh and Cooper (408),
in which the sum of the concentrations of ligand and metal ion
are kept constant and the mole fraction of ligand varied in
steps from 0-1, may be employed. In Job's method the absorbance
at the analytical wavelength is plotted against the mole-fraction
of ligand and the maximum number of ligands bound corresponds to
the point on the abcissa at which absorbance is a maximum or
minimum, depending upon whether the molar absorptivity of the
bound ligand is greater or lower than that of the free ligand
at the analytical wavelength. Once the maximum number of ligands
bound (N) is known, the successive formation constants (k_i) of
the various complexes found between the ligand and the metal ion
may be evaluated from graphical or analytical treatment of the
relationships (411),

$$\bar{n} = \frac{C_L - [L]}{C_M} \qquad\qquad (4.1)$$

and

$$\bar{n} = \frac{\displaystyle\sum_{i=1}^{N} i\beta_i [L]^i}{i + \displaystyle\sum_{i=1}^{N} \beta_i [L]^i} \tag{4.2}$$

where C_L and C_M are the formal concentrations of ligand and metal, respectively, [L] and $\bar{n}$ are the equilibrium concentration of free ligand and the average number of ligands bound to each metal ion at any point in the titration of L with M (or M with L) and $\beta_i = \prod_i k_i$. It should be mentioned here, that the accuracy and precision of the aforementioned treatment by absorptiometry depends very much upon adequate differences between the absorptivities of free and bound ligands at the analytical wavelength (one should be at least 50% greater than the other).

The fluorimetric determinations of the maximum stoichiometry and binding constants of metal-ligand systems follow closely the methodology of the absorptiometric approach. In the fluorimetric approach it is necessary for either the free or the bound ligand or both to fluoresce. If the free and complexed forms of the ligand fluoresce, it is generally necessary that their fluorescence intensities per mole of ligand, at the analytical wavelength, be substantially different. A mole-ratio or Job's type plot in which fluorescence intensity is plotted as the ordinate will allow evaluation of the maximum stoichiometry of the metal-ligand system. In order to calculate the stability constants of the various complexes, however, it is necessary to evaluate the fractions of free and bound ligand and then the absolute concentration of the latter at each point in the titration of the ligand with the metal ion. In the event that only two spectroscopically distinct species are present (i.e. the fluorescence spectra observed are only those of the free ligand and the bound ligand rather than a distinct spectrum for each type of complex),

the total fluorescence intensity measured at the analytical wavelength (I) is the sum of fluorescence intensities of free (I_L)
and bound (I_B) ligand.

$$I = I_L + I_B = k\phi_f \epsilon_L [L] + k\phi_B \epsilon_B [B] \qquad (4.3)$$

In equation (4-3) k is the instrumental constant which is the
product of the excitation intensity the optical path-length and
the factor 2.3, ϕ_L and ϕ_B are the fluorescence-efficiencies, ϵ_L
and ϵ_B, the molar absorptivities and [L] and [B] the equilibrium
concentrations of L and B (the free and bound ligands), respectively. Now, the formal concentration of ligand C_L is given by

$$C_L = [L] + [B] \qquad (4.4)$$

In the absence of metal ion $C_B = [L]$ the fluorescence
intensity is

$$I^O_L = k\phi_L \epsilon_L C_L \qquad (4.5)$$

while in the presence of a sufficient excess of metal ion that
virtually all of the ligand is bound

$$I^O_B = k\phi_B \epsilon_B C_L \qquad (4.6)$$

Equations 4.3, 4.5 and 4.6 are, of course, accurate only if the
absorbance at the wavelength of excitation is less than 0.02.
Otherwise, corrections for the non-linearity of fluorescence
intensity with absorber concentration must be applied. The
combination of equations (4 3) - (4 6) yields the fraction (α_B)
of ligand bound to metal ion.

$$\alpha_B = \frac{I - I^O_L}{I^O_B - I^O_L} \qquad (4.7)$$

The fraction of uncomplexed ligand α_L is then

$$\alpha_L = 1 - \alpha_B = \frac{I^O_B - I}{I^O_B - I^O_L} \qquad (4.8)$$

from which the free ligand concentration can be immediately obtained ($[L] = \alpha_L C_L$) and used in equations (4.1) and (4.2) to obtain the formation constants. Equation (4.8) covers the general case where both free and bound forms of the ligand fluoresce, but with different intensities at the analytical emission wavelengths. However, if only the free ligand fluoresces, equation (4.8) reduces to

$$\alpha_L = \frac{I}{I^O_L} \qquad (4.9)$$

while if the bound ligand is fluorescent but not the free ligand, equation (4.8) becomes

$$\alpha_L = \frac{I^O_B - I}{I^O_B} \qquad (4.10)$$

Equation (4.9) would be most applicable to the complexing of fluorescent ligands by transition metal ions, as the resulting complexes are generally non-fluorescent (some lanthanide and actinide complexes are notable exceptions to this rule). Equations (4.8 and 4.10) would be generally applicable only to non-transition metal complexes and certain lanthanide and actinide chelates in which the line-emission of the metal ion is intensified by chelation.

In light of the low solubilities of many metal chelates, the advantage of fluorimetry (when applicable) over absorptiometry, in their study, is obvious.

The Binding of Small Molecules by Proteins. The binding

of small molecules by proteins is an extremely important process
in biological organisms, as the rates and mechanisms of drug
and metabolite action, distribution and excretion are affected
by protein-binding (413). The binding of drugs and metabolites
to serum proteins is usually a reversible process and, as in the
case of metal-ligand complexation, may be studied, in vitro, as
an equilibrium process. The most extensively studied proteins
in this regard are the low molecular weight ($\sim$ 69,000) bovine
serum albumin and human serum albumin. These proteins possess
relatively intense native fluorescence, by virtue of their
tryptophan residues, in the region around 350 nm. The binding
of small molecules to these proteins is therefore fluorimetrically evaluated either by quenching of the native fluorescence of
the protein (414) or by alteration of the fluorescence of the
ligand upon binding. In regard to the latter phenomenon, the
intense native fluorescence of the proteins makes study of the
ligand fluorescence difficult to evaluate if neither the free
nor bound ligand fluoresces maximally at greater than 400 nm.
The quenching of native protein fluorescence often behaves in a
non-linear fashion during titration with a ligand, possibly due
to conformational changes in the protein during the course of
multiple binding phenomena. The alteration of the fluorescence
of the ligand, however, is usually easier to follow and it is
this process with which the remainder of this discussion will be
concerned.

The determination of the maximum number of ligands bound to
a protein molecule is generally evaluated fluorimetrically by
means of a plot of fluorescence intensity against the mole-ratio
of ligand to protein. This is obtained by titrating the ligand
with an aqueous solution of the protein. Recently, it has been
shown that if the maximum number of ligands bound is small, say
less than 7, Job's method can also be applied, often with greater
accuracy than the mole-ratio method, to the determination of the
maximum stoichiometry (415). Bovine serum albumin and human
serum albumin, in the great majority of cases, appear to bind
from 1 to 5 functionally substituted aromatic ligands, such as

the antimalarial pamaquine of which 3 are bound to bovine serum
albumin (416).

Once the maximum stoichiometry of the protein-ligand complex
has been established, it is in order to evaluate the formation
constants of the various complexes leading to the maximum
stoichiometry. Protein binding equilibria have traditionally
been evaluated by means of the Scatchard equation (417)

$$\frac{\bar{n}}{[L]} = NK - \bar{n}K \qquad\qquad (4.11)$$

or its modified forms (418). In equation (4.11) $\bar{n}$ is the
average number of moles of occupied binding sites (moles of
ligands bound) per mole of protein (as in equation 4.1), and
[L] is the free ligand concentration at any point in the titration of the ligand with a standard solution of the protein. N
is the maximum number of binding sites on the protein (equivalent
to the maximum number of ligands bound) and K is the average
binding constant for the overall binding process. A plot of
$\frac{\bar{n}}{[L]}$ against $\bar{n}$ yields a straight line if all of the binding sites
are non-interacting (i.e. equivalent), whose slope is -K and
whose intercept on the $\frac{\bar{n}}{[L]}$ axis is NK. If more than one type of
binding site for the ligand is present on the protein, equation
(4.11) can be expanded to

$$\bar{n} = \sum_i \frac{K_i N_i [L]}{1 + K_i [L]} \qquad\qquad (4.12)$$

which will give a set of somewhat straight line-segments, each
corresponding to a different N and K, over the range of $\frac{\bar{n}}{[L]}$
and $\bar{n}$ values covered in the titration. If the constants K_i are
well separated (by a factor of 50 or more) each straight line
segment can be extrapolated so that each N_i and each K_i can be
evaluated. However, if the constants are very close, linear
extrapolation will be difficult because of substantial overlap
of the line segments and an analytical treatment, possibly

employing a computer, may be necessary. Because the number of
ligands bound to a protein molecule is usually small, it is
possible to apply the Bjerrum method (411), equations (4.1) and
(4.2), to the binding of small molecules by proteins using the
same approach described for metal-ligand equilibria. In this
case, as in the Scatchard treatment, the free ligand concentration can be determined at any point in the fluorimetric titration
of the ligand with the protein, from the formal ligand concentration and equation (4.8). The constants obtained in this way are
stepwise formation constants rather than an average constant for
all of the binding steps. Moreover, as the Bjerrum treatment
makes no presuppositions as to the nature of the binding sites,
the constants determined from this treatment are independent of
the assumption of the identical nature of the binding sites.
This approach has, along with the Scatchard method, been applied
to the study of the binding to bovine serum albumin of the antimalarial, pamaquine (416). It has also been applied to the
binding of the fluorescent probe, 8-anilino-1-naphthalene-
sulfonate (415), whose binding to serum albumins has been used
for the fluorimetric assay of the proteins (419) to bovine serum
albumin. In both cases the Bjerrum treatment yielded stepwise
formation constants whose respective arithmetic means were in
excellent agreement with the average binding constants evaluated
from the Scatchard treatment.

There is a rather interesting variation on the determination
of the free ligand concentration (equation 4.8) that is applicable
to the fluorimetric study of protein binding but not to that of
metal-ligand binding. Equation (4.8) was derived with the
assumption that the fluorescence intensities of free and bound
ligand were appreciably different at the analytical emission
wavelength. If they were not, binding would not be detectable
fluorimetrically. In many cases, however, it is experimentally
impractical or impossible to choose an analytical emission wavelength at which I^o_L and I^o_B are significantly different. However, in this case, if polarizing prisms or films are placed in
the sample compartment, one in front of the excitation monochromator and one in front of the emission monochromator, with

their optical axes perpendicular to one another, it is possible
to induce differences in the apparent values of I^O_B and I^O_L.
This results from the fact that the free ligands, although
excited by polarized light, in this arrangement, undergo complete rotational depolarization before fluorescing. Thus their
emission intensity is affected only by a small energy loss to
the polarizer components. On the other hand, the bound ligands
retain a significant degree of polarization in their fluorescence by virtue of retardation of their rotational relaxation,
resulting from affixment to the large and slowly rotating protein molecule. Thus, the analyzer part of the polarizing unit,
whose optical axis is perpendicular to a substantial fraction
of the electrical vectors of the light emitted by the bound
ligands, causes much greater attenuation of the fluorescence
of the bound ligands than of that of the free ligands. The use
of polarizers can greatly extend the number of molecules whose
binding to proteins can be studied fluorimetrically.

Up to this point, the experimental aspects of studying
protein binding by fluorimetry were confined to those systems
in which the free ligand was determined in the presence of the
free protein and the bound ligand. When applicable, this
approach is certainly the simplest and fastest. However,
occasionally it is impractical to determine the free ligand
concentration in the presence of free protein and bound ligand.
For example, the fluorescence intensities of the free and bound
ligand are sometimes not sufficiently different, even when the
sample cell is placed between crossed polarizers, to allow determination of the free ligand concentration in the presence of
the bound ligand. In many cases the fluorescences of the free
and bound ligands do not lie sufficiently downfield from that
of the protein and are obscured by the latter. In these cases
fluorimetric determination of the free ligand concentration
can often be effected by separating the free ligand from the
protein and protein-ligand complexes, under equilibrium or
near equilibrium conditions, prior to fluorimetric measurement.
The techniques most often employed for the separation of the

small ligand molecules from the larte proteins and protein-
ligand complexes are equilibrium dialysis, ultracentrifugation
and gel filtration.

Perhaps the most widely employed technique for studying
protein-binding is equilibrium dialysis. In this method, a
semipermeable (usually nitrocellulose) membrane initially
separates an inner buffered protein and an outer protein-free
buffered ligand solution. After the establishment of equilibrium, the free drug concentration in the outer (protein-free)
solution is determined fluorimetrically. The concentration of
bound ligand can, of course, be calculated from the difference
between the formal concentration of ligand added and the free
ligand concentration. The major disadvantage of this method
is the relatively long periods of time required for the establishment of the dialysis equilibrium. During this period un-
stable compounds may decompose or fungal growth may occur. In
many cases, ligand molecules will bind to the dialysis membrane
as well as to the protein. This makes the employment of protein-
free control experiments a necessary part of the equilibrium
dialysis experiment.

Ultracentrifugation (420) entails the preparation of the
drug-protein solution and the subsequent ultracentrifugation
of the sample. The protein-free supernate, containing only the
free ligand is collected from the sample tube and assayed
fluorimetrically. Ultracentrifugation is less time-consuming
and can be carried out with smaller volumes, than equilibrium
dialysis.

Protein-free filtrates can also be obtained by gel-filtration (or membrane ultrafiltration), a relatively recent development. Beads of cross-linked polysaccharides which are water
insoluble but which undergo swelling when in contact with water
are used, in this method, to separate the ligand protein complex
from the free ligand which, because of its smaller size, pene-
trates into the internal volume of the gel-matrix (421). The

successful application of this technique to protein binding
studies requires that the equilibrium between bound and free
drug be slowly reversible. Gel-filtration has been employed,
for example, to study the binding of salicylate to human serum
albumin (422). Both the free and bound forms of salicylate
fluoresce intensely, but with the same fluorescence efficiencies.
In this case, fluorimetric analysis without prior separation
would have been impossible.

The Binding of Small Molecules by Nucleic Acids. Many
substances which function biologically as antimicrobials, anti-
neoplastic agents and mutagens as well as some carcinogens bind
reversibly to the nucleic acids DNA and RNA (423). The vast
majority of these compounds are positively charged at physiological pH and likely interact with the polyanionic nucleic
acids, at least in part, by ionic attraction. It is probably
safe to conclude that virtually any cation, in vivo, is capable
of reversibly binding to some extent, with DNA or RNA. Of the
antimicrobials, antineoplastics and mutagens whose interactions
with DNA and RNA are most often studied (e.g. quinacrine,
chloroquine, proflavin, primaquine, doxorubicin, actinomycin D)
the great majority are derived from polycyclic aromatic systems
which fluoresce in either the free or bound forms, or both.
Although, ionic attraction is present in almost all nucleic
acid-ligand binding, several distinct modes of complexation are
believed to occur (424,425). Because of the variability of the
polymer chain lengths of the nucleic acids and the large numbers
of ligands (> 1000) bound by each nucleic acid strand the
concentrations of nucleic acid solutions are expressed in terms
of component phosphate concentrations (e.g. moles of nucleic
acid phosphate per liter). These concentrations can easily be
calculated from the percentage of phosphorus in the nucleic
acid sample, as supplied by the manufacturer, or they may be
determined spectrophotometrically. In double helical DNA in
which the ratio of aromatic ligand concentration to nucleic
acid concentration is 0.25 or less, a form of binding, known
as intercalation is believed to occur. This process entails

the insertion of the flat aromatic nucleus between adjacent
base pairs of the double helix, accompanied by a 12°-18° un-
winding of the helix in the binding locality. Absorptiometrically or fluorimetrically determined mole-ratio plots demonstrate
that, at least for the larger polycyclic aromatic systems (e.g.
the acridine derivatives), intercalative binding is saturated
when there is one ligand for every 4 or 5 DNA phosphates (424).
At higher ratios of ligand to DNA phosphate external binding
occurs on the double helix and may involve one or even two
ligands per phosphate group. The latter stoichiometry is believed to entail dimerization of the ligand molecules bound to
the outside of the double helix. RNA is capable of demonstrating
the external binding phenomena seen in DNA but does not exhibit
intercalative binding, although there is some evidence to
indicate that a binding akin to intercalation occurs when ligand
molecules are entrapped between single bases of single-stranded
DNA or RNA.

The fluorimetric study of the binding of aromatic ligands
by nucleic acids entails the determination of the stoichiometry
in each class of binding. This is facilitated by the fact that
the fluorescence spectra of free ligand, intercalated ligand
and externally bound ligand are usually appreciably different
from each other either in intensity, position in the spectrum
or both. The determination of the maximum number of ligands
bound in each class of binding is usually carried out by means
of the mole-ratio method, although there is no reason why Job's
method cannot be applied, possibly with greater accuracy, since
the bound ligand to phosphate ratio in each class of binding is
small.

The fluorimetric determination of the average formation
constants in each class of binding is generally effected by
carrying out a titration of a solution of the ligand ($\sim 10^{-5}$-
10^{-6}M) with a concentrated ($\sim 10^{-2}$-10^{-3}M) solution of nucleic
acid (phosphate). The titration data are evaluated graphically
by Peacock and Skerrett's (426) adaptation of the Scatchard
equation (equation 4.11 or 4.12) in which $\bar{n}$ is the average

number of ligands bound per phosphate, K is the average binding
constant but N is now the maximum number of ligands bound per
phosphate (i.e. the number of binding sites per phosphate) in
each class of binding site. [L] is, as usual, the free ligand
concentration. In cases where the fluorescence of the free
ligand is impossible to measure accurately, either because of
overlap with the fluorescences of bound species or because of
low absorber concentration, the equilibrium dialysis, ultracentrifugation or gel-permeation methods are often used to
separate the free ligand and determine its concentration unambiguously.

As in the case of protein binding the slope of each straight
line segment of the Scatchard type plot gives -K for each type
of binding and the $\frac{\bar{n}}{[L]}$ of each extrapolated straight line segment
yields NK for each class of binding. For low values of $\bar{n}$ ($\bar{n} <$
0.25) [L] is very small and $\frac{\bar{n}}{[L]}$ is very large. This region of
the binding isotherm has a very steep slope and therefore a
large binding constant, characteristic of the intercalative mode
of binding. For larger values of $\bar{n}$ ($0.25 < \bar{n} < 2$), [L] is
larger and $\frac{\bar{n}}{[D]}$ smaller than the intercalative region. This
corresponds to the region of external binding and is associated
with binding constants usually an order of magnitude or two
smaller than the constants of the intercalative process.

It has been often observed that in the larger polycyclic
ligands, such as acridine orange (425), aggregation of the free
ligand occurs at relatively low concentrations ($\sim 10^{-6}$-10^{-5}M),
resulting in polymeric species. This is bothersome in the
evaluation of nucleic acid binding, because the free ligand
concentration cannot be accurately determined at higher concentrations without taking into account the polymerization
equilibrium in solution. Aggregation in solution can be avoided
only if the concentration of free ligand is kept below say, 10^{-6}
M. Although nucleic acid binding of aromatic ligands can often
be studied by a variety of physicochemical methods, the great

sensitivity of fluorimetry is virtually indispensible in the
study of the binding of large polycyclic ligands to nucleic
acids.

Covalent Fluorescent-Labeling of Proteins. The
reactions of high fluorescence efficiency, long-wavelength
emitting fluorophores, possessing reactive functional groups,
with proteins, lead to conjugated proteins whose fluorescent
properties are predominately those of the fluorophore. Sul-
phonyl chlorides such as 5-dimethylamino-1-naphthalene-sulphonyl
chloride, isocyanates and isothiocyanates such as rhodamine B-
isocyanate and fluorescein isothiocyanate and diazonium compounds such as rosamine diazonium chloride all yield intensely
fluorescing derivatives when conjugated with proteins (427).
The reactive entities on the proteins which are conjugated with
the prosthetic moieties of the fluorophores are amino, carboxyl
guanidine (arginine) and phenolic (tyrosine) groups. The un-
reacted, hydrolyzed fluorochromes are separated from the labeled
proteins by dialysis or gel filtration.

Covalently labeled fluorescent protein-conjugates are
useful for a variety of biophysical studies such as the probing
of the protein environment through the alteration of the fluorescent properties of the bound fluorophore relative to its
fluorescent properties in solution (428) and the measurement
of intra-protein distances by means of singlet-singlet energy
transfer (429). However, one of the most exciting applications
of covalent labeling is in the area of immunofluorescent staining. It is well known that antibodies, which are generally the
immunoglobulin type proteins, possess a great deal of reactive
specificity for given antigens. The antigens are themselves
proteins associated with bacteria, viruses, parasitic protozoa
and, in general, protein substances foreign to the host organism.
Quite remarkably, the covalent fluorescent labeling of antibodies
has very little effect upon their immunological behavior. Consequently, if a labeled antibody is injected into the host
organism, the localization of the antibody in regions where its

specific antigen is located permits the visualization and location of the invader by fluorescence microscopy of tissue sections
or serum samples. This technique is currently an active area of
research and is currently in routine clinical use in such procedures as the fluorescent treponemal antibody test in the
diagnosis of syphilis and the determination of the antithyroid
activity of serum in the diagnosis of autoimmune diseases (427).
Although conventional fluorescence microscopy allows essentially
qualitative evaluation of the immunofluorescent reaction, the
recent appearance of microspectrofluorimetry promises to put
immunofluorescence on a more quantitative basis (430) and is
certain to make it one of the most powerful of bioanalytical
methods.

TABLE 4.1

Fluorimetric Methods of Analysis for Some Organic Compounds

a.   Direct Methods

Compound	Solvent	pH	λ_{ex} (nm)	λ_{em} (nm)	Sensitivity	Reference
Acenaphthane	pentane	--	291	341	good	5a
Acridine	CF_3COOH	--	358	475	good	84
Allylmorphine	water	1	285	355	poor	5a
p-Aminobenzoic acid	water	8	295	345	fair	5a
Aminopterin	water	7	280 370	460	fair	5a
1-Aminopyrene	CF_3COOH	--	330 342	415	good	84
p-Aminosalicylic acid	water	11	300	405	good	5a
Amobarbital	water	14	265	410	fair	85
Anilines	water	7	280 291	344 361	fair	4
Anthanthrene	pentane	--	350	398	good	5a
Anthracene	pentane	--	420	430	good	5a

Compound	Solvent	pH	λ_{ex}(nm)	λ_{em}(nm)	Sensitivity	Reference
Anthranitic acid	water	2.7	335	422	fair	86
Antimycin A	water	8	350	420	fair	87
Aromatic aldehydes	CH_3OH	--	See Reference	See Reference	fair	88
Ayapin	water	1	350 365	430	good	89
Azaindoles	water	10	290 299	310 347	good	90
Azoguanine	water	7	285	405	poor	5a
Benza-acridines	pentane	--	See Reference	See Reference	good	91
Benzanthrone	CF_3COOH	--	370 420	550	good	84
Benzo(c) acridine	CF_3COOH	--	295 380	480	good	84
Benzo(a) anthracane	pentane	--	284	382	good	5a
Benzo(b) chrysene	pentane	--	283	398	good	5a
11-H Benzo(a) fluorene	pentane	--	317	340	good	5a
Benzoic acid	70% H_2SO_4	--	285	385	good	3
Benzo(e) pyrene	pentane	--	329	389	good	5a

Compound	Solvent	pH	λ_{ex} (nm)	λ_{em} (nm)	Sensitivity	Reference
Benzoquinoline	CF$_3$COOH	--	280	425	good	84
Benzoxanthane	pentane	--	363	418	good	5a
Bromo-lysergic acid diethylamide	water	1	315	460	fair	5a
Brucine	water	7	305	500	fair	5a
Bufotenine	water	<0	292	520	fair	4
Carbazole	dimethyl-formamide	--	291	359	good	92
Chrysene	pentane	--	264	381	good	5a
Cinchonidine	water	1	315	445	poor	5a
Cinchonine	water	1	320	420	poor	5a
Codeine	water	7	285	350	poor	93
Deserpidine	water	1-2	280	365	fair	94
Desipramine	water	~14	295	415	fair	95,96
Dibenzo(a,c)anthracene	pentane	--	280	381	good	5a
Dibenzo(b,k)chrysene	pentane	--	308	428	good	5a
Dibenzo(a,e)pyrene	pentane	--	370	401	good	5a

Compound	Solvent	pH	λ_{ex} (nm)	λ_{em} (nm)	Sensitivity	Reference
5,12-Dihydro-naphthacene	pentane	--	282	340	fair	5a
Diphenyl	water	7	270	318	poor	97
1,4-Diphenyl-butadiene	pentane	--	328	370	good	5a
Epinephrine	water	7	285	325	poor	5a
Esenletin	water	10	365 390	465	fair	89
Ethacridine	water	2	370 425	515	good	98
Fluoranthrene	pentane	--	354	464	good	5a
Fluorene	pentane	--	300	321	good	5a
Gentisic acid	water	7	315	440	fair	5a
Griseofulvin	water	7	295 335	450	good	5a
Harmine	water	1	300 365	400	good	5a
Hippuric acid	70% H_2SO_4	--	270	370	good	3
Hydroxyamphetamine	water	1	275	300	poor	5a
Imipramine	water	14	295	415	fair	96

Compound	Solvent	pH	λ_{ex} (nm)	λ_{em} (nm)	Sensitivity	Reference
Indoleacetic acid	water	8	295	345	fair	5a
Indoles	water	7	269 315	~350	fair	4
Indomethacine	water	13	300	410	good	99
Levo-dromoran	water	1	275	320	poor	5a
Lysergic acid diethylamide	acid	--	325	445	good	26,27,100,101
Menadione	C_2H_5OH	--	335	480	fair	5a
Mephenesin	water	1	280	315	poor	5a
9-Methylanthracene	pentane	--	382	410	good	5a
3-Methylcholanthrene	pentane	--	297	392	good	5a
7-Methyldibenzopyrene	pentane	--	460	467	good	5a
2-Methylphenanthrene	pentane	--	257	357	good	5a
3-Methylphenanthrene	pentane	--	292	368	good	5a
1-Methylpyrene	pentane	--	336	394	good	5a
4-Methylpyrene	pentane	--	338	386	good	5a
Morphine	water	7	285	350	poor	93

Compound	Solvent	pH	λ_{ex} (nm)	λ_{em} (nm)	Sensitivity	Reference
Naphthaleneacetamide	water	11	270 305	327	poor	5a
Naphthalene acetic acid	water	11	270 305	327	poor	5a
Neocinchophen	water	1	275 345	455	poor	5a
Norepinephrine	water	7	285	325	poor	5a
Pamaquine	acidic solution	<0	370	530	fair	5a
Pentobarbital	water	13	265	440	poor	5a
Phenanthrene	pentane	--	252	362	fair	5a
Phenobarbital	water	13	265	440	poor	5a
o-Phenylenepyrene	pentane	--	360	506	good	5a
Phenylenepyrene	water	1	270	305	fair	5a
Physostigmine	water	7	265	315	poor	4
Picene	pentane	--	281	398	good	5a
Piperonyl butoxide	CH_3OH	--	282 302	318	poor	5a
Piperoxan	water	7	290	325	poor	5a

Compound	Solvent	pH	λ_{ex} (nm)	λ_{em} (nm)	Sensitivity	Reference
Podophyllotoxin	water	11	280	325	poor	5a
Procaine	water	11	275	345	fair	5a
Procainamide	water	11	295	385	poor	5a
n-Propylisome	CH3OH	--	280 305	326	poor	5a
Psilocin	water	7	292	314	poor	4
Pyrene	pentane	--	330	382	good	5a
Pyridoxal	water	12	310	365	fair	102
Quinacrine	water	11	285	420	good	5a
Quinidine	water	1	350	450	good	5a
Quinine	water	1	250 350	450	good	5a
Rescinnamine	water	1	310	400	poor	5a
Reserpine	water	1	300	375	good	25
Rutin	acidic solution	--	430	520	fair	5a
Salicylic acid	water	10	310	400	good	5a,63
Scoparone	water	10	350 365	430	good	89

Compound	Solvent	pH	λ_{ex} (nm)	λ_{em} (nm)	Sensitivity	Reference
Scopoletin	water	10	365 390	460	good	89
Streptomycin	water	13	366	445	fair	103
Synephrin	water	1	270	310	good	5a
p-Terphenyl	pentane	--	284	338	good	5a
Thiamylal	water	13	310	530	poor	5a
Thiopental	water	13	315	530	poor	5a
Tribenzo(a,e,i)-pyrene	pentane	--	384	448	good	5a
Triphenylene	pentane	--	288	357	fair	5a
Vitamin A	n-butanol	--	340	490	fair	104
Warfarin	CH$_3$OH	--	290 342	385	fair	5a,105
Yohimbine	water	1	270	360	fair	5a
Zoxazolamine	water	11	280	320	fair	5a

*Sensitivity: good, <0.01 ppm; fair, 0.01 to 0.1 ppm; poor, >0.1 ppm.

TABLE 4.1

b. Indirect Methods

Compound	Reagent and Reaction	Final Solvent or pH of Aq. Solution	λ_{ex} nm	λ_{em} nm	Sensitivity	Reference
Acetylcholine	reduction of ethanol and subsequent esti- mation of unreduced alcohol by alcohol dehydrogenase and NAD	7	340	460	good	106
Acetylfluphenazine	oxidation with H_2O_2	50% CH_3COOH	350	405	fair	32
Acetylsalicylic acid	hydrolysis to salicylic acid	10	313	442	good	62,63,64
Acrolein	reaction with 6-amino- 1-naphthol-3-sulfonic acid	<0	470	500	good	107
Alloxan	condensation with o- phenylenediamine	7	405	520	fair	5
n-Allylnormorphine	heating with con. H_2SO_4	10	365	420	good	36
Amprolium	alkaline oxidation with $AgNO_3$ and $K_3Fe(CN)_6$	n-butanol	400	455	fair	108
Antazoline	reaction with CNBr	7	350	410	poor	109

222 S. G. Schulman

Compound	Reagent and Reaction	Final Solvent or pH of Aq. Solution	λ_{ex} nm	λ_{em} nm	Sensitivity	Reference
Benzoquinolizines	dehydrogenation with mercuric acetate	water	See Ref.		good	110
Bromoridazine	oxidation with H_2O_2	50% CH_3COOH	340	380	poor	32
Carbinoxamine	reaction with CNBr	7	275	467	poor	109
Carphenazine	oxidation with H_2O_2	acid	370	475	poor	32
Chlorcyclizine	oxidation with H_2O_2	7	360	440	poor	32
Chloridazine	oxidation with H_2O_2	acid	340	380	poor	32
Chlorophenothiazine	oxidation with H_2O_2	acid	360	440	poor	32
Chlorothen	reaction with CNBr	7	345	414	fair	109
Chlorpheniramine	reaction with CNBr	7	280	447	poor	109
Chlorpromazine	oxidation with H_2O_2	acid	340	385	fair	33,99
Chlorpromazine sulfoxide	oxidation with H_2O_2	acid	340	380	fair	32,33
Chlorprothixene	oxidation with H_2O_2	acid	345	410	poor	32
Chlortetracycline	heating with alkali	alkali	350	420	fair	111
Coumaphos	hydrolysis	alkali	330	410	good	65
Cyclizine	oxidation with H_2O_2	7	305	417	poor	109

Compound	Reagent and Reaction	Final Solvent or pH of Aq. Solution	λ_{ex} nm	λ_{em} nm	Sensitivity	Reference
Dexpanthenol	alkaline hydrolysis followed by reaction with ninhydrin	9.1	385	465	fair	112
Digitoxigenin	heating with strong acid	acid	395	570	fair	113
Digitoxin	heating with strong acid	acid	350	490	fair	113
Diphenylhydrazine	oxidation with H_2O_2	7	305 345	412 454	poor	109
Doxylamine	reaction with CNBr	7	280	388	poor	109
Emetine	reaction with iodine in alkali	C_2H_5OH	436	570 620	fair	114
Epinephrine	oxidation to tri-hydroxyindole	3.5	436	540	good	5a
Epinephrine	oxidation to tri-hydroxyindole	3.5	365	490	good	95
Fluphenazine	oxidation with H_2O_2	acid	350	405	fair	32
Formaldehyde	reaction with acetyl-acetone and ammonia	6	410	510	good	115
Gitoxigenin	heating with strong acid	acid	350	465	fair	5a

Compound	Reagent and Reaction	Final Solvent or pH of Aq. Solution	λ_{ex} nm	λ_{em} nm	Sensitivity	Reference
Gitoxin	heating with strong acid	acid	350	470	fair	113
Glutathione	condensation with o-phthaldehyde	8	343	425	good	116,117
Guthion	alkaline hydrolysis to anthranilic acid	Benzene	340	400	good	118
Heroin	heating with con. H_2SO_4	10	365	420	good	36
Hydrocortisone	heating with ethanol and con. H_2SO_4	con. acid	470	525	good	119
Hydrogen peroxide	oxidation of diacetyl-dichlorofluorescein in presence of peroxidase	7.2	490	530	good	120
2-Imidazole dihydro-morphanthridine	heating with acetic acid	acid	310	410	fair	121
Isoniazid	condensation with salicylaldehyde followed by reduction	iso-butanol	392	478	good	35,122
Malic acid	reaction with resorcinol	8.5	490	530	fair	123
Malonaldehyde	condensation with 4,4-sulfonyldianiline	acidic dimethylformamide	490	550	good	60

Compound	Reagent and Reaction	Final Solvent or pH of Aq. Solution	λ_{ex} nm	λ_{em} nm	Sensitivity	Reference
Maretin	alkaline hydrolysis	Benzene	372	480	good	124
Meclizine	oxidation with H_2O_4	7	310 345	420 444	poor	109
6-Mercaptopurine	oxidation with $KMnO_4$	alkali	305	400	fair	125
Mescaline	condensation with ammonia and formaldehyde	acid	375	575	good	126
Metanephrine	oxidation to tri-hydroxyindole	7	425	525	good	127
Metaraminol	condensation with o-phthaldehyde	acid	370	500	good	128
Methapyrilene	reaction with CNBr	7	350	412	good	109
Methiomeprazine	oxidation with H_2O_2	acid	360	440	fair	32
Methiophenothiazine	oxidation with H_2O_2	acid	360	385	good	32
Methotrexate	oxidation with $KMnO_4$	5	280 370	450	fair	5a
Methoxypromazine	oxidation with H_2O_2	acid	340	380	poor	32
Methyldylidenyl-thixin	oxidation with H_2O_2	acid	280	370	poor	32

S. G. Schulman

Compound	Reagent and Reaction	Final Solvent or pH of Aq. Solution	λ_{ex} nm	λ_{em} nm	Sensitivity	Reference
Methyltrifluoopera-zine	oxidation with H_2O_2	acid	350	455	fair	32
Morphine	heating with con. H_2SO_4	10	365	420	good	36,129
Norchlorpromazine	oxidation with H_2O_2	acid	350	385	poor	32
Norchlorpromazine	oxidation with H_2O_2	acid	340	375	poor	32
Norespinephrine	oxidation to tri-hydroxyindole	6.5	395	490	good	5a,130
Oxalic Acid	reduction followed by reaction with resor-cinol	~12	490	530	good	131
Oxytetracycline	complexing with Mg^{2+} and versene	10.6	390	520	fair	68
Penicillin	reaction with 2-methoxy-6-chloro-9-β-amino-ethylacridine	acid	420	500	fair	5a
Perphenazine	oxidation with H_2O_2	acid	345	380	poor	32
Pheniramine	reaction with CNBr	7	275	434	poor	109
α-Phthalic acid	reaction with resorcinol	13	490	530	poor	132
Potasan	hydrolysis	13	490	515	good	5a

Compound	Reagent and Reaction	Final Solvent or pH of Aq. Solution	λ_{ex} nm	λ_{em} nm	Sensitivity	Reference
Prochlorperazine	oxidation with H_2O_2	acid	340	380	poor	32
Promazine	oxidation with H_2O_2	acid	340	375	fair	99
Promethazine	oxidation with H_2O_2	acid	340	375	poor	32
Pyridoxine	reaction with KCN	9.5	358	435	good	133
Pyridylchlorophenothiazine	oxidation with H_2O_2	acid	330	550	poor	32
Pyrilamine	reaction with CNBr	7	350	419	fair	109
Reserpine	reaction with p-toluene-sulfonic acid in glacial acetic acid	CH_3COOH	380	480	good	134
Streptomycin	reaction with β-naphtho-quinone-4-sulfonate	alkali	~365	445	good	135
Sulfonamides	reaction with 4,5-methylenedioxy-phthaldehyde	acid	320 327	375 425	good	136
Tetracycline	complexing with Ca^{2+} and barbituric acid	~8	405	530	good	70
Thenyldiamine	reaction with CNBr	7	355	414	good	109

S. G. Schulman

Compound	Reagent and Reaction	Final Solvent or pH of Aq. Solution	λ_{ex} nm	λ_{em} nm	Sensitivity	Reference
Thiethylperazine	oxidation with H_2O_2	50% CH_3COOH	360	445	fair	32
Thioridazine	oxidation with H_2O_2	50% CH_3COOH	365	440	good	33,99
Thioridazine disulfone	oxidation with H_2O_2	50% CH_3COOH	360	440	poor	32
Thioridazine disulfoxide	oxidation with H_2O_2	50% CH_3COOH	360	435	poor	32
Thioridazine-R-sulfoxide	oxidation with H_2O_2	50% CH_3COOH	360	440	poor	32
Thioridazine-S-sulfoxide	oxidation with H_2O_2	50% CH_3COOH	360	435	poor	32
Thioridothixin	oxidation with H_2O_2	50% CH_3COOH	320	395	poor	32
Trifluormeprazine	oxidation with H_2O_2	50% CH_3COOH	350	405	fair	32
Trifluorophenothiazine	oxidation with H_2O_2	50% CH_3COOH	350	410	fair	32
Trifluoropromazine	oxidation with H_2O_2	50% CH_3COOH	350	405	fair	32
Tripelennamine	reaction with CNBr	7	355	419	fair	32
d-Tubocurarine	complexation with the dye rose bengal	acetone	570	590	good	74,75

Compound	Reagent and Reaction	Final Solvent or pH of Aq. Solution	λ_{ex} nm	λ_{em} nm	Sensitivity	Reference
Vitamin C	condensation with o-phenylenediamine	∿9	350	430	fair	137
Vitamin D	reaction with acetic anhydride $-H_2SO_4$	acid	390	470	good	138

*Sensitivity: good, <0.01 ppm; fair, 0.01 to 0.1 ppm; poor, >0.1 ppm.

TABLE 4.2

Phosphorescence Data for Some Organic Compounds and for Compounds of Biological Interest

Compound	Solvent[a]	Wavelengths[b] λ_{ex} (nm) λ_{em} (nm)		Lifetime[c] sec	Limit of Detection µg/ml	Reference
Acenaphthene	EtOH	300	515	---	0.2	139
Acetaldehyde-4-nitro-phenylhydrazone	EPA	395	525	0.5	0.06	140
Acetone-4-nitrophenyl-hydrazone	EPA	392	525	0.48	0.1	140
3-Acetylpyridine	EtOH	277	424	<0.5	3.6	153
N-Acetyl-L-tyrosine ethyl ester	WME	250	395	---	0.1	154
Adenine	WM	278	406	2.9	0.02	155
Adenosine	EtOH	280	422	0.8	3.2	153
p-Aminobenzoic acid	EtOH	305	425	---	0.001	156
2-Aminofluorene	EtOH	380	590	4.6	0.01	20
6-Amino-6-methyl-mercaptopurine	WM	321	456	0.66	0.0002	156
2-Amino-4-methyl-pyrimidine	EtOH	302	438	2.1	0.033	153

Compound	Solvent[a]	Wavelengths[b] λ_ex (nm)	λ_em (nm)	Lifetime[c] sec	Limit of Detection μg/ml	Reference
2-Amino-5-nitro-benzothiazole	EPA	375	515	0.41	0.08	140
2-Amino-5-nitro-biphenyl	EPA	380	520	0.56	0.05	140
L-3-Aminotyrosine·2HCl	EtOH	286	398	0.8	2.4	153
Anabasine	EtOH	270	390	6.2	0.01	29,142
Anthracene	EtOH	300	462	---	0.05	139
Apomorphine·HCl	EtOH	320	470	3.1	0.001	143
Aramite	EtOH	285	400	3.3	0.0003	29,142
L-Arterenol-bitartrate	EtOH	260	455	0.5	1.0	20
Aspirin	EPA	240	380	2.1	0.10	28,144
Atropine	EtOH	d	410	1.4	0.10	145
8-Azaguanine	EtOH	282	442	1.8	0.3	153
Bayer 44646[e]	EPA	290	460	0.6	0.01	142,146
Bayer 37344[e]	EPA	275	435	<0.2	0.01	142,146
Benzaldehyde	EtOH	254	433	3.4	0.004	20
1,2-Benzanthracene	EtOH	310	510	2.2	0.03	139

Compound	Solvent[a]	Wavelengths[b] λ_{ex} (nm)	λ_{em} (nm)	Lifetime[c] sec	Limit of Detection µg/ml	Reference
1,2-Benzfluorene	EtOH	315	502	---	0.2	139
2,3-Benzfluorene	EtOH	325	502	---	0.2	139
Benzimidazole	EtOH	280	406	2.3	0.006	153
Benzocaine	EtOH	310	430	3.4	0.007	139
Benzoic acid	EPA	240	400	2.4	0.005	145
Benzophenone-4-nitro-phenylhydrazone	EPA	365	515	---	2.0	140
4-Benzolybiphenyl-4-nitrophenylhydrazone	EPA	370	520	0.38	0.4	140
3,4-Benzpyrene	EtOH	325	508	---	3.0	139
Benzyl alcohol	EtOH	219	393	---	0.04	20
6-Benzylaminopurine	WM	286	413	2.8	0.02	155
Biphenyl	EtOH	270	385	1.0	0.004	147
6-Bromopurine	WM	273	420	0.5	0.002	155
Brucine	EtOH	305	435	0.9	0.1	143
Bufotenine mono-oxalate	WMH	301	447	1.9	0.001	154

Compound	Solvent[a]	Wavelengths[b] λex(nm)	λem(nm)	Lifetime[c] sec	Limit of Detection μg/ml	Reference
Butacaine sulfate	EtOH	310	430	5.7	0.05	145
Caffeine	EtOH	285	440	2.0	0.2	145
Carbazole	EtOH	341	436	7.8	0.001	157
2-Chloro-4-amino-benzoic acid	EtOH	312	447	1.0	0.07	153
Chlorobenzilate	EtOH	275	415	<0.2	0.001	142,146
p-Chlorophenol[e]	EtOH	290	505	<0.2	0.02	142,146
o-Chlorophenoxy-acetic acid	EtOH	280	518	0.7	0.2	158
p-Chlorophenoxy-acetic acid	EtOH	283	396	<0.5	0.004	158
6-Chloropurine	WM	273	419	0.64	0.002	155
Chlorpromazine·HCl	EtOH	320	490	0.3	0.03	145
Chlortetracycline	EtOH	280	410	2.7	0.05	145
Cincophen	EtOH	350	520	0.8	0.02	145
Cocaine·HCl	EtOH	240	400	2.7	0.01	145
Codeine	EtOH	270	505	0.3	0.01	143

Compound	Solvent[a]	Wavelengths[b] λ_{ex}(nm)	λ_{em}(nm)	Lifetime[c] sec	Limit of Detection µg/ml	Reference
Co-Ral[e]	EtOH	335	510	<0.2	0.00004	142,146
Cyclaine·HCl	EtOH	240	400	2.4	0.006	145
Cytidine[f]	WM	287 290 295	412 420 415	---	2.9	159
Cytidine[f]	WM2	292 291 295	420 420 424	---	0.016	159
DDD (p,p[1])[e]	EtOH	265	415	<0.2	0.001	142,146
DDE (p,p[1])[e]	EtOH	270	425	0.2	0.0002	142
DDT (p,p[1])[e]	EtOH	270	420	0.2	0.007	142
Desoxypyridoxine·HCl	EtOH	290	442	1.4	0.076	153
Diacetylsulfanilanide	EtOH	280	405	1.3	0.001	148
2,6-Diaminopurine sulfate	EtOH	294	424	1.7	1.2	153
2,6-Diaminopurine	WM	288	410	2.7	0.15	158
Diazinon	EtOH	275	395	5.0	0.03	142,146
1,2,5,6-Dibenzan-thracene	EtOH	340	550	1.3	0.003	139

Compound	Solvent[a]	Wavelengths[b] λ_{ex} (nm)	λ_{em} (nm)	Lifetime[c] sec	Limit of Detection μg/ml	Reference
2,6-Dichloro-4-nitro-aniline	EPA	368	525	0.47	0.04	140
2,4-Dichlorophenoxy-acetic acid		289	490	<0.5	0.002	158
Dicumarol	EtOH	305	475	0.6	0.001	149
2,6-Diethyl-4-nitro-aniline	EPA	388	525	0.66	0.19	140
3,4-Dihydroxymardelic acid[f]	EtOH	260 294 293	397 412 415	<0.5 1.1 1.05	0.09	160
3,4-Dihydroxyphenyl-acetic acid[f]	EtOH	267 295 292	390 430 428	<0.5 0.9 0.9	0.2	160
2,5-Dimethoxy-4-methyl-amphetamine	WM	289	411	3.9	0.01	154
5,7-Dimethyl-1,2-benzaacridine	EtOH	310	555	0.6	0.03	20
N,N-Dimethyl-4-nitro-aniline	EPA	398	525	0.54	0.05	140
N,N-Dimethyltryptamine	WM	286	434	6.9	0.015	154

Compound	Solvent[a]	Wavelengths[b] λ_{ex}(nm)	λ_{em}(nm)	Lifetime[c] sec	Limit of Detection μg/ml	Reference
Diphenadione	EtOH	260	440	0.6	1.0	20
Dopa[f]	EtOH	286 292 293	435 427 424	0.9 0.9 1.3	0.1	160
Dopamine[f]	EtOH	273 285 293	410 430 436	0.9	0.15	160
Ephedrine	EtOH	225	390	3.6	0.20	145
Epinephrine[f]	EtOH	260 283 281	412 425 429	0.6 1.0 0.8	0.2	160
L-Epinephrine bitar- trate	EtOH	270	410	0.4	1.0	20
Estradiol	EtOH	292	403	2.0	0.3	153
N-Ethylcarbazole	EtOH	340	437	7.8	0.001	157
N-Ethylcarbazole	CHX	298	433	8.1	0.001	157
Ethyl-3-indoleacetate	EtOH	290	440	3.3	0.02	158
Folic acid	EtOH	367	425	---	0.004	156
Guthion[e]	EtOH	325	420	0.6	0.06	142,146

Compound	Solvent[a]	Wavelengths[b] λ_{ex}(nm)	λ_{em}(nm)	Lifetime[c] sec	Limit of Detection μg/ml	Reference
Hippuric acid	EPA	311	450	4.9	0.004	20
Homovanillic acid[f]	EtOH	292 279 282	418 435 439	1.1 0.8 0.8	0.04	160
DL-5-Hydroxytryptophan	EtOH	315	435	6.3	0.1	20
Ibocaine·HCl	WM	292	430	8.6	0.01	154
Imidan[e]	EtOH	305	440	0.75	0.0006	142,146
Indole-3-acetic acid	EtOH	290	438	<0.5	0.02	158
3-Indoleacetonitrile	EtOH	285	438	7.1	0.005	20
3-Indolebutyric acid	EtOH	284	510	0.6	0.004	158
Indolecarboxylic acid	EtOH	290	429	5.5	0.0006	20
3-Indolepropionic acid	EtOH	290	440	0.6	0.002	158
Isolan[e]	EtOH	285	395	1.6	0.2	142,146
Kelthane[e]	EtOH	285	515	<0.2	0.0006	142,146
Kepone[e]	EtOH	260	410	1.2	1.0	142,146
Lidocaine	EtOH	265	400	1.1	1.2	145
D-Lysergic acid	WM	310	518	0.1	---	154

Compound	Solvent[a]	Wavelengths[b] λ_{ex}(nm)	λ_{em}(nm)	Lifetime[c] sec	Limit of Detection μg/ml	Reference
Lysergic acid diethyl-amide (LSD)	WM3	311	512	0.004	0.008	154
Mebaral	EtOH	240	380	2.2	0.01	145
Metanephrine[f]	EtOH	288 280 277	418 432 432	1.3 1.1 1.1	0.02	160
Methoxychlor[e]	EtOH	275	380	0.7	0.0004	142,146
3-Methoxy-4-hydroxy-phenylethylamine[f]	EtOH	294 286 284	423 440 440	1.2 <0.5 0.7	0.1	160
Methycaine·HCl	EtOH	240	400	2.7	0.006	145
2-Methylcarbazole	EtOH	333	442	8.1	0.001	157
2-Methylcarbazole	CHX	332	443	7.5	0.001	157
N-Methylcarbazole	EtOH	336	437	8.4	0.001	157
N-Methylcarbazole	CHX	298	431	7.5	0.001	157
6-Methylmercaptopurine	WM	291	420	0.6	0.006	155
N-Methyl-4-nitroaniline	EPA	390	522	0.5	0.05	140
6-Methylpurine	WM	272	405	3.2	0.01	155

Compound	Solvent[a]	Wavelengths[b] λ_{ex}(nm)	λ_{em}(nm)	Lifetime[c] sec	Limit of Detection μg/ml	Reference
Morphine	EtOH	285	500	0.3	0.01	143
Morphine sulfate	EtOH	265	460	0.8	10.0	143
Naphthacene	EtOH	300	518	---	0.001	139
Naphthalene	EPA	310	475	1.8	0.7	150
α-Naphthalene acetamide	EtOH	297	513	2.5	0.004	158
Naphthaleneacetic acid	EtOH	295	510	2.8	0.0004	158
α-Naphthol	EtOH	320	475	1.15	0.0002	142,146
β-Naphthoxyacetic acid	EtOH	328	497	2.6	0.006	158
β-Naphthylamine	EtOH	270	505	2.3	0.03	20
Narceine	EtOH	290	440	0.5	0.1	143
NIA 10242[e]	EtOH	285	400	1.6	0.007	142,146
Niacinamide	EtOH	270	410	---	0.2	156
Nicotine	EtOH	270	390	5.2	0.01	146,151
5-Nitroacenaphthene	EPA	380	540	---	0.5	140
4-Nitroaniline	EPA	380	510	0.6	0.02	140
9-Nitroanthracene	EPA	248	488	---	0.13	140

Compound	Solvent[a]	Wavelengths[b] λ_{ex} (nm) λ_{em} (nm)		Lifetime[c] sec	Limit of Detection μg/ml	Reference
1-Nitroanthraquinone	EPA	250	490	0.28	0.25	140
4-Nitrobiphenyl	EPA	330	480	---	0.2	140
2-Nitrofluorene	EPA	340	517	0.40	0.04	140
6-Nitroindole	EPA	372	520	0.41	0.08	140
1-Nitronaphthalene	EPA	340	520	---	1.5	140
2-Nitronaphthalene	EPA	260	500	0.36	0.15	140
4-Nitro-1-naphthylamine	EPA	400	578	---	60	140
3-Nitro-N-ethylcarbazole	EPA	315	475	0.37	0.01	140
2-Nitro-N-methylcarbazole	EPA	345	530	---	2.8	140
4-Nitro-2-toluidine	EPA	375	520	0.53	0.1	140
4-Nitrophenylhydrazine	EPA	390	520	0.48	0.03	140
4-Nitrophenol	EtOH[g]	355	520	<0.2	0.00002	29,146
Norepinephrine[f]	EtOH	284 292 292	386 425 430	0.5 1.2 1.2	0.15	160
Normetanephrine[f]	EtOH	290 283 280	416 435 435	0.65 0.6 0.6	0.02	160

Compound	Solvent[a]	Wavelengths[b] λ_{ex} (nm)	λ_{em} (nm)	Lifetime[c] sec	Limit of Detection μg/ml	Reference
Nornicotine	EtOH	270	390	5.3	0.01	20,151
Orthotran[e]	EtOH	260	395	<0.2	0.002	142,146
Oxythiamine·HCl	EtOH	272	460	<0.5	3.4	153
Papavarine·HCl	EtOH	260	480	1.5	0.0005	143
Parathion[e]	EtOH	360	515	<0.2	0.008	142,146
Phenacetin	EPA	d	410	---	0.2	157
Phenanthrene	EPA	340	465	2.6	1.0	152
Phencyclidine (PCP)	WM	265	385	---	0.32	161
Phenindione	EtOH	235	395	<0.2	1.0	149
Phenobarbital	EtOH	240	380	1.8	0.1	145
Phenylalanine	EtOH	270	385	---	0.4	20
Phenylephrine·HCl	EtOH	290	390	2.4	0.01	145
DL-β-Phenyllactic acid	EtOH	262	383	5.4	5.0	153
Phthalylsulfacetamide	EtOH	290	415	0.6	0.001	148
Phthalylsulfathiazole	EtOH	305	405	0.9	1.0	148
α-Piolinic acid·HCl	EtOH	278	400	5.2	0.014	153

Compound	Solvent[a]	Wavelengths[b] λ_{ex}(nm)	λ_{em}(nm)	Lifetime[c] sec	Limit of Detection μg/ml	Reference
Procaine·HCl	EtOH	310	430	3.5	0.01	145
Propionaldehyde-4-nitrophenylhydrazone	EPA	395	525	0.5	0.06	140
Psilocin	WM	292	443	4.0	0.006	154
Psilocybin	WM	280	430	5.5	0.014	154
Purine	WM	272	405	2.2	0.01	155
Pyrene	EtOH	329	515	---	0.2	20
Pyridine	EtOH	310	440	1.4	0.0001	148
Pyridine-3-sulfonic acid	EtOH	272	408	1.2	4.8	153
Pyridoxine·HCl	EtOH	291	425	---	0.008	156
Quercetin	EtOH	343	480	2.1	0.3	153
Quinidine sulfate	EtOH	340	500	1.3	0.04	145
Quinine·HCl	EtOH	340	500	1.3	0.04	145
Retene	EtOH	265	510	---	0.001	139
Ronnel[e]	EtOH	300	475	<0.2	0.0006	142,146
Rutonal	EtOH	240	380	2.5	0.2	142,146

Compound	Solvent[a]	Wavelengths[b] λ_{ex}(nm)	λ_{em}(nm)	Lifetime[c] sec	Limit of Detection μg/ml	Reference
Salicylic acid	EtOH	315	430	6.2	0.05	20
Serotonin	EtOH	d	410	---	50	144
Sevin[e]	EtOH	300	510	2.0	0.004	142,146
Sodium sulfathiazole	EtOH	315	410	1.4	1.0	148
Strychnine phosphate	EtOH	290	440	1.2	0.05	143
Sulfabenzamide	EtOH	305	405	0.7	0.0001	151
Sulfacetamide	EtOH	280	410	1.3	0.0001	151
Sulfadiazine	EtOH	275	410	0.7	0.01	151
Sulfaguamide	EtOH	305	405	0.7	0.01	151
Sulfamerazine	EtOH	280	405	0.7	0.0001	151
Sulfamethazine	EtOH	280	410	0.8	0.0001	151
Sulfanilamide	EtOH	297	411	2.9	0.012	153
Sulfapyridine	EtOH	310	440	1.4	0.0001	141,151
Sulfathiazole	EtOH	310	420	0.9	1.0	151
Sulfenone	EtOH	275	390	<0.2	0.0005	151
Tedion[e]	EtOH	295	410	<0.2	0.0002	142,146

Compound	Solvent[a]	Wavelengths[b] λ_{ex}(nm)	λ_{em}(nm)	Lifetime[c] sec	Limit of Detection μg/ml	Reference
1,2,4,5-Tetramethyl-benzene	EPA	275	392	4.5	1.8	152
Thebaine	EtOH	315	500	1.0	1.0	143
2-Thiouracil	EtOH	312	432	<0.5	0.0038	153
α-Tocopherol	EtOH	296	430	---	0.05	156
2-Tolidine	EtOH	310	510	2.2	0.02	20
Toxaphene[e]	EtOH	240	390	1.9	0.02	142,146
2,4,5-Trichlorophenol	EtOH	305	485	<0.2	0.003	142,146
2,4,5-Trichlorophenoxy acetic acid	EtOH	294	473	1.1	0.0005	158
2,4,5-Trichlorophenoxy propionic acid	EtOH	294	467	<0.5	0.003	158
Triphenylene	EtOH	291	461	15	0.0002	20
Trithion[e]	EtOH	305	430	<0.2	0.003	142,146
Tromexan	EtOH	295	460	0.6	0.01	149
Tronothane·HCl	EtOH	300	410	1.2	0.02	145
Tryptophan	EtOH	295	440	1.5	0.002	141,20

Compound	Solvent[a]	Wavelengths[b] λ_{ex}(nm) λ_{em}(nm)		Lifetime[c] sec	Limit of Detection μg/ml	Reference
Tyrosine[f]	EtOH	253 291 290	394 390 389	1.9 2.8 3.3	0.02	160
U.C. 10854[e]	EtOH	270	385	2.9	0.002	142,146
Vitamin K$_1$	EtOH	348	564	0.65	1.0	162
Vitamin K$_1$	NHX	346	570	0.4	0.5	162
Vitamin K$_3$	MeOH	339	546	0.7	0.4	162
Vitamin K$_3$	EtOH	338	545	0.76	0.4	162
Vitamin K$_3$	NHX	335	508	0.54	0.1	162
Vitamin K$_3$	WM	338	546	0.6	0.15	162
Vitamin K$_3$	WM4	342	542	0.6	0.07	162
Vitamin K$_3$	WM5	340	545	---	0.08	162
Vitamine K$_5$	WM	310	535	1.3	0.1	162
Vitamine K$_5$	WM4	317	530	1.35	0.08	162
Warfarin	EtOH	305	460	0.8	0.01	149
Yohimbine·HCl	EtOH	290	410	7.4	0.01	143
Zectran[e]	EtOH	285	440	0.45	0.005	142,146

Mordant Blue 9
Lumogallion
8-Hydroxyquinoline
Benzoin
1-Amino-4-hydroxy
Alizarin Red S
Acetylsalicyl
Rhodamine B
Kojic ac
Butyl

Boron (B)

Gold (Au)

a Solvent abbreviations: EtOH = ethyl alcohol; EPA = 5:5:2 volume ratio of diethyl ether, isopentene, and ethyl alcohol; WM = 9:1 volume ratio of water and methyl alcohol; WME = 5:11:4 volume ratio of water, methyl alcohol, and ethyl alcohol; WM1 = 0.2 $\underline{M}$ NaOH in WM; WM2 = 0.1 $\underline{M}$ NaI in WM; WM3 = 0.75 $\underline{M}$ NaI in WM; CHX = cyclohexane; NHX = n-hexane; MeOH = methyl alcohol; WM4 = 2:8 volume ratio of water and methyl alcohol; WM5 = 3:7 volume ratio of water and methyl alcohol.

b Only peak wavelengths are given.

c Decay times are given for peak wavelengths.

d No filter or monochromator was used for the exciting radiation.

e These compounds are pesticides.

f Order of peak wavelengths are for basic, neutral, acidic solutions.

g Solvent made basic with 2 drops of diethylamine per 10 ml of ethyl alcohol.

Ion	Method of Estimation	Reagent	Limit of Detection
Aluminum (Al) (Continued)	C	Morin	
	C	Pontachrome BBR	
	C	Pontachrome VSW	
	C	Quercetin	
	C	N-Salicylidene-2-a... hydroxyfluorene	
	Chem	Gutzeit test	
	C	Uranyl nit...	
	Rd	Cerium(...)	
Arsenic (As)			

Compound	Solvent[a]	Wavelengths[b] λ_{ex} (nm) λ_{em} (nm)		Lifetime[c] sec	Limit of Detection µg/ml	Reference
Tyrosine[f]	EtOH	253 291 290	394 390 389	1.9 2.8 3.3	0.02	160
U.C. 10854[e]	EtOH	270	385	2.9	0.002	142,146
Vitamin K₁	EtOH	348	564	0.65	1.0	162
Vitamin K₁	NHX	346	570	0.4	0.5	162
Vitamin K₃	MeOH	339	546	0.7	0.4	162
Vitamin K₃	EtOH	338	545	0.76	0.4	162
Vitamin K₃	NHX	335	508	0.54	0.1	162
Vitamin K₃	WM	338	546	0.6	0.15	162
Vitamin K₃	WM4	342	542	0.6	0.07	162
Vitamin K₃	WM5	340	545	---	0.08	162
Vitamine K₅	WM	310	535	1.3	0.1	162
Vitamine K₅	WM4	317	530	1.35	0.08	162
Warfarin	EtOH	305	460	0.8	0.01	149
Yohimbine·HCl	EtOH	290	410	7.4	0.01	143
Zectran[e]	EtOH	285	440	0.45	0.005	142,146

a Solvent abbreviations: EtOH = ethyl alcohol; EPA = 5:5:2 volume ratio of diethyl ether, isopentene, and ethyl alcohol; WM = 9:1 volume ratio of water and methyl alcohol; WME = 5:11:4 volume ratio of water, methyl alcohol, and ethyl alcohol; WM1 = 0.2 $\underline{M}$ NaOH in WM; WM2 = 0.1 $\underline{M}$ NaI in WM; WM3 = 0.75 $\underline{M}$ NaI in WM; CHX = cyclohexane; NHX = n-hexane; MeOH = methyl alcohol; WM4 = 2:8 volume ratio of water and methyl alcohol; WM5 = 3:7 volume ratio of water and methyl alcohol.

b Only peak wavelengths are given.

c Decay times are given for peak wavelengths.

d No filter or monochromator was used for the exciting radiation.

e These compounds are pesticides.

f Order of peak wavelengths are for basic, neutral, acidic solutions.

g Solvent made basic with 2 drops of diethylamine per 10 ml of ethyl alcohol.

TABLE 4.3

Indirect Fluorimetric Methods of Analysis for Some Inorganic Ions

Ion	Method of Estimation	Reagent	Limit of Detection (ppm)	Reference
Silver (Ag)	C	Butylrhodamine S	0.01	194
	Q	Eosin + 1,10-phenanthroline	0.004	195
	C	8-Hydroxyquinoline-5-sulfonic acid	0.013	196
	Cat	Lucigenin + H_2O_2	0.08	197
	Q	1H-Naphtho[2,3-d]triazole	0.1	198
	C	Resorufin	0.01	199
Aluminum (Al)	C	Acid Alizarin Garnet R	0.007	200
	C	Coumarin derivatives	0.20	201
	C	Flazo Orange	0.001	202
	C	3-Hydroxy-2-naphthoic acid	0.01	203,204
	C	8-Hydroxyquinoline	0.10	205
	C	Lumogallion	0.014	206
	C	Mordant Blue 9	0.005	207

Ion	Method of Estimation	Reagent	Limit of Detection (ppm)	Reference
Aluminum (Al) (Continued)	c	Morin	0.05	208,209
	c	Pontachrome BBR	0.02	210
	c	Pontachrome VSW	0.02	210
	c	Quercetin	0.05	211
	c	N-Salicylidene-2-amino-3-hydroxyfluorene	0.001	212–214
Arsenic (As)	Chem	Gutzeit test	1.0	180
	c	Uranyl nitrate	100	216–218
	Rd	Cerium(IV)	7.5	220
Gold (Au)	c	Butylrhodamine B	0.1	220
	c	Kojic acid	0.1	221
	c	Rhodamine B	0.02	222
	c	Acetylsalicylic acid	0.01	223
Boron (B)	c	Alizarin Red S	1.0	224
	c	1-Amino-4-hydroxy-anthraquinone	1.0	225
	c	Benzoin	0.04	226–230

Ion	Method of Estimation	Reagent	Limit of Detection (ppm)	Reference
Boron (B) Continued	C	Cochineal Red	1.0	231
	C	Dibenzoylmethane	0.0005	232
	C	Flavonol	1.0	233
	C	Morin	1.0	21
	C	Phenylfluorone	1.0	234
	C	Quercetin	1.0	235
	C	Quinalizarin	0.01	236
	C	Resacetophenone	1.0	237,238
	C	Rhodamine 6G + salicylic acid	0.001	239
	C	Thoron I	0.005	240
Barium (Ba)	In	Curcumin (Turmeric Yellow)	20	216,241
	C	Fluorexone	80.0	242,243
Beryllium (Be)	C	1-Amino-4-hydroxyanthraquinone	0.2	244,245
	C	Benzoin	0.1	226
	C	1,4-Dihydroxyanthraquinone	0.2	245
	C	2-(2'-Hydroxyphenyl)benzothiazole	0.1	246

Ion	Method of Estimation	Reagent	Limit of Detection (ppm)	Reference
Beryllium (Be) (Continued)	C	8-Hydroxyquinaldine	0.001	247
	C	Morin	0.01	216-218 245,248-251
	C	Substituted 2-hydroxy-3-naphthoic acid	0.09	252
	C	Tetracycline + 5,5-diethyl-2-thiobarbituric acid	0.10	253
Bromide (Br⁻)	Chem	Fluorescein	1.0	254
	Q	Uranyl nitrate	1.0	255
Calcium (Ca)	C	Calcein	0.2	256-260
	In	Curcumin	2.0	216,241
	C	Fluorexone	1.0	242,243
	C	8-Hydroxyquinoline	1.0	241,261
	C	8-Quinolylhydrazone	0.2	262
Cadmium (Cd)	C	2-(2'-Hydroxyphenyl)-benzoxazole	2	263
	C	8-Hydroxyquinoline	2	265
	C	Morin	2	265

Ion	Method of Estimation	Reagent	Limit of Detection (ppm)	Reference
Cadmium (Cd) (Continued)	C	p-Tosyl-8-aminoquinoline	0.02	266
Cerium (Ce)	C	Sulfonaphtholazoresorcinol	0.05	267
Chloride (Cl⁻)	Q	Uranyl nitrate	1.0	255
Cyanide (CN⁻)	C	Chloramine T + nicotinamide	0.3	268
	Q	Pd complex of 8-hydroxy-quino-line-5-sulfonic acid	0.02	269
	C	Quinone	0.01	271,272
	Cat	Pyridoxal	0.02	270
Cobalt (Co)	Q	Al-Pontachrome BBR	0.001	273
Chromium (Cr)	Q	Triazinylstilbexone	0.004	274
Copper (Cu)	C	Cochineal Red	2.0	216,241
	Q	2-(2'-Hydroxyphenyl)benzoxazole	0.1	275
	Q	1-(2-Hydroxypropyl)anabasine	0.05	276
	C	Luminocupferron	0.1	277
	C	Rose Bengal Extra + 1,10-phenanthroline	0.1	278

Ion	Method of Estimation	Reagent	Limit of Detection (ppm)	Reference
Copper (Cu) (Continued)	C	Salicylalazine	0.05	279
	C	Thiamine	0.1	280
	C	1,1,3-Tricyano-2-amino-1-propene	0.1	281
Cesium (Cs)	C	8-Hydroxyquinoline	0.1	282,283
Europium (Eu)	C	Benzoyltrifluoroacetone	0.003	284,285
	C	Hexafluoroacetone-trioctyl phosphine oxide	0.0001	286
	C	Tetradentate complex with 2-theonyltrifluoroacetone, collidine, and diphenyl-guanidine	1.0	288
	C	2-Theonyltrifluoroacetone	0.0001	289
Fluoride (F⁻)	Q	Al-Acid Alizarin Garnet R complex	0.001	200
	Q	Al-morin complex	0.2	290
	Q	Mg-8-hydroxyquinoline complex	0.01	291
	C	Ternary complex with Zr + Calcein Blue	0.01	292
	Q	Zr-3-hydroxyflavone complex	0.1	293

Ion	Method of Estimation	Reagent	Limit of Detection (ppm)	Reference
Iron (Fe)	Q	Cochineal Red	1.0	216,241
	Q	α-naphthoflavone	1.0	216,241
	Q	Rhodamine S	1.0	294
	Q	2,2',6',2'-Terpyridyl	0.01	295
Gallium (Ga)	C	5,7-Dibromo-8-hydroxyquinoline	0.1	296
	C	1-(2,4-Dihydroxyphenylazo)-2-naphthol-4-sulfonic acid	0.01	297
	C	8-Hydroxyquinaldine	0.02	298,299
	C	8-Hydroxyquinoline	0.05	300,301
	C	Lumogallion	0.1	302-304
	C	Morin	0.1	265,305,306
	C	2-(2'-Pyridyl)benzimidazole	0.07	307
	C	Rhodamine B	0.01	308,309
	C	Rhodamine 6G	0.1	310
	C	Salicylidene-o-aminophenol	0.1	311
	C	Solochrome Red ERS, Black AS, or 6BFA	0.01	308,309

Ion	Method of Estimation	Reagent	Limit of Detection (ppm)	Reference
Gallium (Ga) (Continued)	C	Sulfonaphtholazoresorcinol 2,2'-4'-Trihydroxy-5-chloro-1,1'-azobenzene-3-sulfonic acid	0.001	316
Germanium (Ge)	C	Benzoin	2.0	317
	C	Resacetophenone	100	238
	C	Trihydroxyanthraquinone	2.0	318
Hafnium (Hf)	C	Flavonol	0.1	319
	C	Quercetin	1.0	320
Mercury (Hg)	Q	Rhodamine B	0.1	321
Iodide (I⁻)	Q	Luminol	1.0	322
	Q	α-Naphthoflavone	1.0	216-218
	Q	Uranyl nitrate	1.0	255
	Rd	Cerium(IV)	0.6	219
Indium (In)	C	8-Hydroxyquinaldine	0.2	323
	C	8-Hydroxyquinoline	0.04	324
	C	Morin	0.2	325
	C	2-(2'-Pyridyl)benzimidazole	0.1	307

Ion	Method of Estimation	Reagent	Limit of Detection (ppm)	Reference
Indium (In) (Continued)	C	Pyronine Y	5.0	326
Iridium (Ir)	C	2,2',2" -Terpyridyl	2.0	327
Potassium (K)	C	8-Hydroxyquinoline	1.0	261,282,328
	C	Zinc Uranyl acetate	1.0	218
Lithium (Li)	C	Dibenzothiazolylmethane	0.5	329
	C	8-Hydroxyquinoline	0.1	330
	C	Quercetin	1.0	331
	C	Uranyl nitrate	1.0	217
Magnesium (Mg)	C	Fluoran	0.01	332,333
	C	8-Hydroxyquinoline	0.01	334-338
	C	1-(8-hydroxyquinoline-7-azo)-2-naphthol-4-sulfonic acid	0.01	339
	C	1-(2-Hydroxy-3-sulfo-5-chloro-phenylazo)-2'-hydroxynaphthalene	0.020	340
	C	Lumomagneson	0.004	341
	C	Bis(salicylideneamino)benzofuran	0.010	342

Ion	Method of Estimation	Reagent	Limit of Detection (ppm)	Reference
Magnesium (Mg) (Continued)	C	Bissalicylideneethylenediamine	0.0002	343
Manganese (Mn)	C	8-Hydroxyquinoline-5-sulfonic acid	0.005	344
Molybdenum (Mo)	C	Carminic acid	0.9	345,346
	C	8-Hydroxyquinoline sodium-tetraphenylborate	0.2	347
	C	Primuline	20	348
Ammonium (NH_4^+)	C	Hantzch reaction	0.01	349,350
	Ez	NADH	0.01	351
Nitrate (NO_3^-)	C	2,3-Diaminonaphthalene	0.01	352
Sodium (Na)	C	8-Hydroxyquinoline	1.0	282,283
	Q	Zinc uranyl acetate	1.0	353
Nickel (Ni)	Q	Al-1-(2-pyridylazo)-2-naphthol	0.00003	354
Osmium (Os)	Rd	Cerium(IV)	0.5	219
Oxygen (O_2)	Oxd	Acriflavine	0.01	355-358
	Oxd	9,10-Dihydroacridine	0.01	359

Ion	Method of Estimation	Reagent	Limit of Detection (ppm)	Reference
Oxygen (O_2) (Continued)	Oxd	Epinephrine	2.0	360
	Oxd	Fluorescein	0.01	361
Ozone (O_3)	Oxd	9,10-Dihydroacridine	0.01	362,363
	C	2-Diphenylacetyl-1,3-indandione-1-hydrazone	0.02	364
Phosphate (PO_4^{3-})	Q	Al-morin	0.05	365
	C	Molybdophosphate-Rhodamine B	0.04	366
	Ez	NADPH	0.01	367
Lead (Pb)	C	Morin	5.0	218
Rubidium (Rb)	C	8-Hydroxyquinoline	5.0	180
Ruthenium (Ru)	C	5-Methyl-1,10-phenanthroline	1.0	368
Sulfide (S^{2-})	C	Fluorescein mercuriacetate	0.00005	369
	Q	Pd complex with 8-hydroxy-quinoline-5-sulfonic acid	0.2	269
Sulfate (SO_4^{2-})	Q	Th-morin	24	370
Antimony (Sb)	C	Luminol	0.05	371
	C	Rhodamine 6G	0.1	372

Ion	Method of Estimation	Reagent	Limit of Detection (ppm)	Reference
Antimony (Sb) (Continued)	C	2,4',7-Trihydroxyflavone	0.04	373
Scandium (Sc)	C	5,7-Dichloro-8-hydroxyquinoline	0.1	374
	C	Morin + phenazone	0.01	375
	C	Salicylalsemicarbazide	1.0	376
Selenium (Se)	C	3,3'-Diaminobenzidine	0.02	377–379
	C	2,3-Diaminonaphthalene	0.02	380–385
Silicon (Si)	C	Ammonium molybdate	0.003	386
Samarium (Sm)	C	Hexafluoroacetone-trioctyl-phosphine oxide	0.1	286
	C	1,10-Phenanthroline + 2-phenyl-cinchoninic acid ternary complex	0.5	387
	C	2-Theonyltrifluoroacetone	0.0001	289
	C	2-Theonyltrifluoroacetone ternary complex	10.0	288
Tin (Sn)	C	Flavonol	0.1	388
	C	8-Hydroxyquinoline-5-sulfonic acid	0.005	389

Ion	Method of Estimation	Reagent	Limit of Detection (ppm)	Reference
Tin (Sn) (Continued)	C	Morin	0.2	265
Strontium (Sr)	C	Fluorexone	80.0	242,243
Terbium (Tb)	C	Antipyrine + salicylate	0.1	390
	C	EDTA-sulfosalicylic acid	0.006	391
	C	Hexafluoroacetone-trioctyl-phosphine oxide	0.1	286
	C	4,4'-Methylenedi-[3-methyl-1-(2-pyridyl)pyrazol-5-ol]	0.025	392
	C	Phenyl salicylate	0.1	393
Therium (Th)	C	1-Amino-4-hydroxyanthraquinone	8	228
	C	Flavonol	0.01	394
	C	Morin	0.02	395
	C	Quercetin	0.02	396
Titanium (Ti)	C	Salicylic acid	1.0	218
Thallium (Tl)	Q	Cochineal Red	1.0	218
	C	Rhodamine B	0.1	397,398

Ion	Method of Estimation	Reagent	Limit of Detection (ppm)	Reference
Thallium (Tl) (Continued)	Q	Uranyl sulfate	1.0	218
Uranium (U)	Q	Morin	0.05	399
Vanadium (V)	C	Resorcinol	2.5	400
Tungsten (W)	C	Carminic acid	0.3	345,346
	C	3-Hydroxyflavone	1.0	401
	Q	Rhodamine B	1.0	402
Yttrium (Y)	C	5,7-Dibromo-8-hydroxyquinoline	0.1	403
	C	8-Hydroxyquinoline	0.02	404
Zinc (Zn)	C	Benzoin	0.5	405
	C	8-Hydroxyquinoline	1.0	406,407
	C	Luminocupferron	0.2	408
	C	2,2'-Methylenedibenzothiazole	2.0	409
	C	Picolinaldehyde-2-quinolyl-hydrazone	0.026	410
	C	p-Tosyl-8-aminoquinoline	0.02	266
Zirconium (Zr)	C	Flavonol	0.1	319

Ion	Method of Estimation	Reagent	Limit of Detection (ppm)	Reference
Zirconium (Zr) (Continued)	C	Morin	0.02	411
	C	2,4',7-Trihydroxyflavone	0.05	412

*Methods of estimation: C, chelation; Cat, catalytic; Chem, chemical; Ez, enzymatic; In, indicator; Rd, reduction; Oxd, oxidation; Q, quenching.

REFERENCES

1. De Ment, J., _Fluorochemistry_, Chemical Publishing, Brooklyn, 1945.

2. Brederek, K., Forster, T. and Oesterlin, H. G., in _Luminescence of Organic and Inorganic Materials_, (H. P. Kallman and G. M. Spruch, Eds.), Wiley, New York, 1962, p. 161.

3. Ellman, G. L., Burkhalter, A. and La Dou, J., _J. Lab. Clin. Med._, 57, 813 (1961).

4. Bridges, J. W. and Williams, R. T., _Biochem. J._, 107, 225 (1968).

5. (a) Udenfriend, S., _Fluorescence Assay in Biology and Medicine_, Academic Press, New York, N. Y., 1962.

   (b) Udenfriend, S., _Fluorescence Assay in Biology and Medicine_, Volume II, Academic Press, New York, N. Y., 1969.

6. Cowgill, R. W., _Arch. Biochem. Biophys._, 100, 36 (1963).

7. Brodie, B. B., Udenfriend, S., Dill, W. and Downing, G., _J. Biol. Chem._, 168, 311 (1947).

8. Schenk, G. and Wirz, D., _Anal. Chem._, 42, 1754 (1970).

9. Pringsheim, P., _Fluorescence and Phosphorescence_, Interscience, New York, N. Y., 1949.

10. D. M. Hercules, Ed., Fluorescence and Phosphorescence Analysis, Interscience, New York, N. Y., 1966.

11. White, C. E. and Weissler, A., in _Handbook of Analytical Chemistry_, (L. Meites, Ed.), McGraw Hill, New York, N. Y., 1963, Chapter 6.

12. White, C. E. and Weissler, A., in _Standard Methods of Chemical Analysis_, Volume IIIA (F. J. Welcher, Ed.), Van Nostrand, Princeton, New Jersey, 1966, Chapter 5.

13. White, C. E. and Weissler, A., _Anal. Chem._, 36, 116R (1964); 38, 115R (1966); 40, 114R (1968); 42, 57R (1970); 44, 182R (1972).

14. Weissler, A., _Anal. Chem._, 46, 500R (1974).

15. Berlman, I. B., _Handbook of Fluorescence Spectra of Aromatic Molecules_, Academic Press, New York, N. Y., 1965.

16. Phillips, R. E. and Elevitch, E. R., in Progress in Clinical Pathology (M. Steffani, Ed.), Grune and Stratton, New York, N. Y., 1966, Chapter 4.

17. Passwater, R. A., Guide to Fluorescence Literature, Plenum Press, New York, N. Y., 1967.

18. Konstantinova-Shlesinger, M. A., Fluorimetric Analysis (N. Kaner, Transl.), Davey, New York, N. Y., 1965.

19. Bartos, J. and Pesez, M., Talanta, 19, 93 (1972).

20. Winefordner, J. D., St. John, P. A. and McCarthy, W. J., "Phosphorimetry as a Means of Chemical Analysis", Chapter 2 in reference 5b.

21. White, C. E. and Argauer, R. J., Fluorescence Analysis. A Practical Approach, Marcel Dekker, New York, N. Y., 1970.

22. Winefordner, J. D., Schulman, S. G. and O'Haver, T. C., Luminescence Spectroscopy in Analytical Chemistry, Wiley-Interscience, New York, N. Y., 1972.

23. Smith, H. F., Res. Develop., July, 20 (1968).

24. (a) Guilbault, G. G., in Fluorescence, Theory, Instrumentation and Practice (G. G. Guilbault, Ed.), Marcel Dekker, New York, N. Y., 1967.

24. (b) Guilbault, G. G., Practical Fluorescence - Theory, Methods, and Techniques, Marcel Dekker, New York, N. Y., 1973.

25. Udenfriend, S., Duggan, D. E., Vasta, B. M. and Brodie, B. B., J. Pharmacol. Exptl. Therap., 120, 26 (1957).

26. Axelrod, J., Bradly, R. O., Witkop, B. and Evarts, E. V., Ann. N. Y. Acad. Sci., 66, 435 (1957).

27. Aghajanian, G. K. and Bing, O. H. L., Clin. Pharmacol. Therap., 5, 611 (1964).

28. Winefordner, J. D. and Latz, H. W., Anal. Chem., 35, 1517 (1963).

29. Moye, H. A. and Winefordner, J. D., J. Agr. Food Chem., 13, 533 (1965).

30. Mellinger, T. J. and Keeler, C. E., Anal. Chem., 35, 554 (1963).

31. Mellinger, T. J. and Keeler, C. E., Anal. Chem., 36, 1840 (1964).

32. Ragland, J. B. and Kinross-Wright, V. J., Anal. Chem., 36, 1357 (1964).

33. Ragland, J. B., Kinross-Wright, V. J. and Ragland, R. S.,
    Anal. Biochem., 12, 60 (1965).

34. Pearlman, E. J., J. Pharmacol. Exptl. Therap., 95, 465 (1949).

35. Peters, J. H., Am. Rev. Respirat. Diseases, 81, 485 (1960).

36. Nadeau, G. and Sobolewski, G., Can. J. Biochem. Physiol.,
    36, 625 (1958).

37. Balatre, P. H., Traisnel, M. and Delcambre, J. R., Am.
    Pharm. Franc., 19, 171 (1961).

38. Albers, R. W. and Lowry, O. H., Anal. Chem., 27, 1829 (1955).

39. McDougal, Jr., D. B. and Farmer, H. S., J. Lab. Clin. Med.,
    50, 485 (1957).

40. Udenfriend, S., Stein, S., Bohlen, P., Leimgruber, W. and
    Wiegele, M., Science, 178, 871 (1973).

41. Stein, S., Bohlen, P., Imai, K., Stone, J. and Udenfriend,
    S., Fluorescence News, 7, 9 (1973).

42. Wiegele, M., DeBernardo, S. and Leimgruber, W., Biochem.
    Biophys. Res. Commun., 50, 352 (1973).

43. Wiegele, M., DeBernardo, S., Leimgruber, W., Cleeland, R.
    and Grunberg, E., Biochem. Biophys. Res. Commun., 54,
    899 (1973).

44. Felix, A. M. and Jimenez, M. H., Anal. Biochem., 52, 377
    (1973).

45. Narasimhachari, A., Biochem. Biophys. Res. Commun., 55, 231
    (1973).

46. Felix, A. M. and Terkelsen, G., Arch. Biochem. Biophys.,
    157, 177 (1973).

47. Chen, R. F., Anal. Lett., 7, 65 (1974).

48. Felix, A. M., Jiminez, M. H., J. Chromatogr., 89, 361 (1974).

49. Klein, B., Sheehan, J. E. and Grunberg, E., Clin. Chem., 20,
    272 (1974).

50. de Silva, J. A. F. and Strojny, N., Anal. Chem., 47, 714
    (1975).

51. Samejima, K., Dairman, W., Stone, J. and Udenfriend, S.,
    Anal. Biochem., 42, 237 (1971).

52. Lowe, I. P., Robins, E. and Eyerman, G. S., J. Neurochem.,
    3, 8 (1958).

53. McCaman, M. W. and Robins, E. J., _J. Lab. Clin. Med._, _59_, 885 (1962).

54. Wong, P. W. K., O'Flynn, M. E. and Inouyte, T., _Clin. Chem._, _10_, 1098 (1964).

55. Velluz, L., Perez, M. and Herbain, M., _Bull. Soc. Chim. France_, _15_, 681 (1948).

56. Camber, B., _Nature_, _174_, 1107 (1954).

57. Brandt, P. and Cheronis, N., _Microchem. J._, _5_, 110 (1961).

58. Sawicki, E., in _International Symposium on Microchemical Techniques_ (N. Cheronis, Ed.), Wiley, New York, N. Y., 1962, p. 59.

59. Sawicki, E., Stanley, T. and Pfaff, J., _Chemist-Analyst_, _51_, 9 (1962).

60. Sawicki, E., Stanley, T. and Johnson, H., _Anal. Chem._, _35_, 199 (1963).

61. Spikner, J. E. and Towne, J. C., _Anal. Chem._, _34_, 1468 (1962).

62. Cotty, V. F. and Ederma, H. M., _J. Pharm. Sci._, _55_, 837 (1966).

63. Lange, W. E. and Bell, S. A., _J. Pharm. Sci._, _55_, 386 (1966).

64. Harris, P. A. and Riegelman, S., _J. Pharm. Sci._, _56_, 713 (1967).

65. MacDougall, D., _Residue Rev._, _1_, 24 (1962).

66. MacDougall, D., _Residue Rev._, _5_, 119 (1964).

67. Anderson, C. A., Adams, J. M. and MacDougall, D., _J. Agr. Food Chem._, _7_, 256 (1959).

68. Ibsen, K. H., Saunders, R. L. and Urist, M. R., _Anal. Biochem._, _5_, 505 (1963).

69. Kelly, R. G., Floyd, H. A. and Hoyt, K. D., in _Antimicrobial Agents and Chemotherapy_, Am. Soc. Microbiol., Ann Arbor, Michigan, 1966, p. 666.

70. Kohn, K. W., _Anal. Chem._, _33_, 863 (1961).

71. Glazko, A. J., Dill, W. A. and Fransway, R. L., _Federation Proc._, _21_, 269 (1962).

72. Laugel, M. P., _Compt. Rend._, Acad. Sci., _255_, 692 (1962).

73. Ogawa, S., Morita, M., Nishiura, K. and Fujisawa, K., _J. Pharm. Soc. Japan_, (Yakugaku Zasshi), 85, 650 (1965).

74. Cohen, E. N., J. Lab. Clin. Med., 61, 338 (1963).

75. Cohen, E. N., J. Lab. Clin. Med., 62, 979 (1963).

76. Sturgeon, R. J. and Schulman, S. G., Anal. Chim. Acta, 75, 225 (1975).

77. Karush, F., Klinman, N. R. and Marks, R., Anal. Biochem., 9, 100 (1964).

78. Guilbault, G. G., Enzymatic Methods of Analysis, Pergamon Press, Oxford, England, 1970.

79. Purdy, W. C., Electroanalytical Methods in Biochemistry, McGraw-Hill, New York, N. Y., 1965.

80. Guilbault, G. G., Sardar, M. H. and Peres, K., Anal. Biochem., 31, 19 (1969).

81. Guilbault, G. G., Brignac, P. and Zimmer, M., Anal. Chem., 40, 190 (1968).

82. Guilbault, G. G., Brignac, P. and Juneau, M., Anal. Chem., 40, 1256 (1968).

83. Pardue, H., Burke, M. and Jones, D. O., J. Chem. Ed., 44, 684 (1967).

84. Sawicki, E., Stanley, T. W. and Johnson, H., Microchem. J., 8, 257 (1964).

85. Swagzdis, J. E. and Flanagan, T. L., Anal. Biochem., 7, 147 (1964).

86. Nakken, K. F., Scan. J. Clin. Lab. Invest., 15, suppl. 76 (1963).

87. Sehgal, S. N. and Vezina, C., Anal. Biochem., 21, 266 (1967).

88. Crowell, E. P. and Varsel, C. J., Anal. Chem., 35, 189 (1963).

89. Crosby, D. G. and Berthold, R. V., Anal. Biochem., 4, 349 (1962).

90. Adler, T. K., Anal. Chem., 34, 683 (1962).

91. Sawicki, E., Stanley, T. W., Pfaff, J. D. and Elbert, W. C., Anal. Chim. Acta, 31, 359 (1964).

92. Bender, D. F., Sawicki, E. and Wilson, R. M., Anal. Biochem., 36, 1011 (1964).

93. Brandt, R., Ehrlich-Rogozinsky, S. and Cheronis, N. D., Microchem. J., 5, 215 (1961).

94. Gordon, J. A. and Campbell, D. J., Anal. Chem., 29, 488 (1957).

95. Small, N. A., _Clin. Chim. Acta_, <u>8</u>, 803 (1963).

96. Dingell, J. V., Salser, F. and Gillette, J. R., _J. Pharmacol. Exp. Ther._, <u>143</u>, 14 (1964).

97. Bridges, J. W., Creaven, P. J. and Williams, R. T., _Biochem. J._, <u>96</u>, 872 (1965).

98. Issekutz, B. and Hajdu, P., _Arzneim. Forsch._, <u>16</u>, 645 (1966).

99. Tomsett, S. L., _Acta Pharmacol. et Toxicol._, <u>26</u>, 298 (1967).

100. Genist, K. and Farmilo, C. G., _J. Pharm. Pharmacol. Therap._, <u>16</u>, 250 (1964).

101. Dal Cortivo, L. A., Broich, J. R., Dihrberg, A. and Newman, B., _Anal. Chem._, <u>38</u>, 1959 (1966).

102. Bridges, J. W., Davies, D. S. and Williams, R. T., _Biochem. J._, <u>105</u>, 1261 (1967).

103. Faure, F. and Blanquet, P., _Clin. Chim. Acta_, <u>9</u>, 292 (1964).

104. Drujan, D. B., Castillon, R. and Gueroro, E., _Anal. Biochem._, <u>23</u>, 44 (1968).

105. Corn, M. and Berberich, R., _Clin. Chem._, <u>13</u>, 126 (1967).

106. Cooper, J. R., _Biochem. Pharmacol._, <u>13</u>, 795 (1964).

107. Sawicki, E., Stanley, T. W. and Pfaff, J., _Anal. Chim. Acta_, <u>28</u>, 156 (1963).

108. Szalkowski, C. R., _J. Assn. Offic. Agric. Chemists_, 48, 285 (1965).

109. Jensen, R. E. and Pflaum, R. J., _J. Pharm. Sci._, <u>53</u>, 835 (1964).

110. Schwartz, D. E. and Rieder, J., _Clin. Chim. Acta_, <u>6</u>, 453 (1961).

111. Katz, S. E. and Spock, J., _J. Assn. Offic. Agric. Chemists_, <u>47</u>, 203 (1964).

112. Panier, R. G. and Close, J. A., _J. Pharm. Sci._, <u>53</u>, 108 (1964).

113. Wells, D., Katzung, B. and Meyers, F. H., _J. Pharm. Pharmacol._, <u>13</u>, 389 (1961).

114. Ichimura, _Bunseki Kagaku_, <u>10</u>, 623 (1961); _Chem. Abstr._, <u>56</u>, 1530e (1962).

115. Belman, S., _Anal. Chim. Acta_, <u>29</u>, 120 (1963).

116. Cohn, V. H. and Lyle, J., *Anal. Biochem.*, 14, 434 (1966).

117. McNeil, T. L. and Beck, L. V., *Anal. Biochem.*, 22, 431 (1968).

118. Adams, J. M. and Anderson, C. A., *Agric. Food Chem.*, 14, 53 (1966).

119. Mattingley, D., *J. Clin. Path.*, 15, 374 (1962).

120. Keston, A. S. and Brandt, R., *Anal. Biochem.*, 11, 1 (1965).

121. Tishler, F., Hagman, H. E. and Brody, S. M., *Anal. Chem.*, 37, 906 (1965).

122. Scott, E. M. and Wright, R. C., *J. Lab. Clin. Med.*, 70, 333 (1967).

123. Shassman, M., Ceci, L. and Tucci, A. F., *Anal. Biochem.*, 23, 484 (1968).

124. Anderson, R. J., Anderson, G. A. and Yagelowich, M. L., *Agric. Food Chem.*, 14, 43 (1966).

125. Finkel, J. M., *Anal. Biochem.*, 21, 362 (1967).

126. Seiler, N. and Wiechmann, M., *J. Phys. Chem.*, 337, 229 (1964).

127. Kahane, Z. and Vesterguard, P., *J. Lab. Clin. Med.*, 70, 333 (1967).

128. Shone, P. A. and Alpers, H. S., *Life Sci.*, 3, 551 (1964).

129. Kupferberg, H., Burkhalter, A. and Way, L. A., *J. Pharmacol. Exp. Ther.*, 145, 247 (1964).

130. Fleming, R. M., Clark, W. G., Fenster, E. D. and Towne, J. C., *Anal. Chem.*, 37, 692 (1965).

131. Zaremboki, P. M. and Hodgkinson, A., *Biochem. J.*, 96, 717 (1965).

132. Thommes, G. A. and Leininger, E., *Talanta*, 5, 260 (1960).

133. Polansky, M. M., Camarra, R. T. and Toepfer, E. W., *J. Assn. Offic. Agric. Chem.*, 47, 827 (1964).

134. Jakovljevic, I. M., Fose, J. M. and Kuzel, N. R., *Anal. Chem.*, 34, 410 (1962).

135. Issekutz, B. and Hajdu, P., *Arzheim. Forsch.*, 16, 645 (1966).

136. Amano, T., *Yaku. Zasshi*, 85, 1045 (1965).

137. Deutsch, M. J. and Weeks, E. C., *J. Assn. Offic. Agric. Chem.*, 48, 1248 (1965).

138. Chen, P. S., Terepka, A. R. and Lane, K., Anal. Biochem.,
     8, 34 (1964).

139. Hood, L. V. S. and Winefordner, J. D., Anal. Chem., 38,
     1922 (1966).

140. Sawicki, E. and Pfaff, J., Microchem. J., 12, 7 (1967).

141. Lukasiewicz, R. J., Mousa, J. J. and Winefordner, J. D.,
     Anal. Chem., 44, 1339 (1972).

142. Moye, H. A. and Winefordner, J. D., J. Agr. Food Chem.,
     13, 516 (1965).

143. Hollifield, H. C. and Winefordner, J. D., Talanta, 12,
     860 (1965).

144. Latz, H. W., Ph.D. thesis, University of Florida, Gaines-
     ville, Florida, 1963.

145. Winefordner, J. D. and Tin, M., Anal. Chim. Acta, 31, 239
     (1964).

146. Moye, H. A., Ph.D. thesis, University of Florida, Gaines-
     ville, Florida, 1965.

147. McCarthy, W. J. and Winefordner, J. D., J. Assoc. Off. Agr.
     Chemists, 48, 915 (1965).

148. Hollifield, H. C. and Winefordner, J. D., Anal. Chim. Acta,
     36, 352 (1966).

149. Hollifield, H. C. and Winefordner, J. D., Talanta, 14, 103
     (1967).

150. McGlynn, S. P., Daigre, J. and Smith, F. J., J. Chem. Phys.,
     39, 675 (1963).

151. Winefordner, J. D. and Moye, H. A., Anal. Chim. Acta, 32,
     278 (1965).

152. McGlynn, S. P., Neely, B. T. and Neely, W. C., Anal. Chim.
     Acta, 28, 472 (1963).

153. Saunders, L. B., Cetorelli, J. J. and Winefordner, J. D.,
     Talanta, 16, 407 (1969).

154. Aaron, J. J., Spann, W. J. and Winefordner, J. D., Talanta,
     20, 855 (1973).

155. Aaron, J. J. and Winefordner, J. D., Anal. Chem., 44, 2127
     (1972).

156. Aaron, J. J. and Winefordner, J. D., Anal. Chem., 19, 21
     (1972).

157.   Perry, A. W., Tidwell, P., Cetorelli, J. J. and Wineford-
       ner, J. D., Anal. Chem., 43, 781 (1971).

158.   Sanders, L. B. and Winefordner, J. D., J. Agr. Food Chem.,
       20, 166 (1972).

159.   Perry, A. W., Tidwell, P., Cetorelli, J. J. and Winefordner,
       J. D., Anal. Chem., 43, 781 (1971).

160.   Spann, W. J., Mousa, J. J., Aaron, J. J. and Winefordner,
       J. D., Anal. Biochem., 53, 154 (1973).

161.   Fabrick, D. M. and Winefordner, J. D., Talanta, 20, 1220
       (1973).

162.   Aaron, J. J. and Winefordner, J. D., Anal. Chem., 44, 2122
       (1972).

163.   Bozhevol'nov, E. A., Fluorimetric Analysis of Inorganic
       Materials (in Russian), Khimiza, Moscos, 1960.

164.   Randall, J. T., Trans. Faraday Soc., 35, 1 (1939).

165.   Arkangelskaya, V. and Feofilov, P., Opt. Spektrosk., 2,
       107 (1957).

166.   Leveranz, H. W., Luminescence of Solids, Chapman and Hall
       Ltd., London, 1950.

167.   Vavilov, S. and Levshin, V., Z. Physik, 48, 397 (1928).

168.   Price, G., Ferritti, R. and Schwartz, S., Anal. Chem., 28,
       1651 (1956).

169.   Gentappi, F., Ross, A. and Sesa, M., Anal. Chem., 28, 1651
       (1956).

170.   Thatcher, L. and Baker, F., Anal. Chem., 29, 1575 (1957).

171.   Steele, T. W. and Robert, R. V., Nucl. Sci. Abstr., 22,
       2754 (1968).

172.   White, C. E., Anal. Chem., 26, 129 (1954).

173.   White, C. E., Anal. Chem., 24, 87 (1952).

174.   White, C. E., Anal. Chem., 30, 729 (1958).

175.   Zaidel, A. N. and Larionore, Y., Dokl. Akad. Nauk., SSSR,
       16, 443 (1937).

176.   Furukawa, M., Sasaki, S., Nakashima, R. and Shibata, S.,
       Nagoya Kogyo Gijutau Shikensho Hogoku, 17, 251 (1968).

177.   Poluektov, N. S., Kirillov, A., Tishichenko, M. A. and
       Zelyukova, Y., Zh. Anal. Khim., 22, 707 (1967).

178. Fassel, V., Heidel, R. and Huke, R., Anal. Chem., 24, 606
      (1952).

179. Fassel, V. and Heidel, R., Anal. Chem., 26, 1134 (1954).

180. Haitinger, M., Die Fluorszenzanalyse in der Mikrochemie,
      Wien-Leipzig, 1937.

181. Neunhoeffer, O., Z. Anal. Chem., 132, 91 (1951).

182. Belzi, M. U. and Kushnirenko, I., Referat. Zh. Khim., 196D,
      1969, Abstr. # 16G92.

183. Belzi, M. U. and Kushnirenko, I., Referat. Zh. Khim., 166D,
      1969, Abstr. # 6G67.

184. Solov'ev, A. and Bozhevol'nov, E. A., Chem. Abstr., 66,
      B2105N (1967).

185. Kirkbright, G. F. and Saw, C. G., Talanta, 15, 570 (1968).

186. Belzi, M. U. and Kushnirenko, I., Referat. Zh. Khim., 196D,
      1969, Abstr. # 6G66.

187. Kirkbright, G. F., Saw, C. G. and West, T. S., Analyst, 94,
      457 (1969).

188. Shcherbov, D. P. and Ivankova, A. L., Prom. Khim. Reakt.
      Osob. Chist. Veshchestv., 8, 191 (1967).

189. Shcherbov, D. P., Astaf'eva, I. N., Plotnikova, R. N.,
      Issled. Obl. Khim. Fiz. Metod. Anal. Miner. Syr'ya, 18
      (1971); Chem. Abstr., 78, 66448u (1973).

190. Astaf'eva, I. N., Shcherbov, D. P. and Plotnikova, R. N.,
      Issled. Obl. Khim. Fiz. Metod. Anal. Miner. Syr'ya , 30
      (1971); Chem. Abstr., 78, 66554a (1973).

191. Brand, J. C. D., Humphrey, D. R., Douglas, A. E. and Zanon,
      I., Can. J. Phys., 51, 530 (1973).

192. Okabe, H., Splitstone, P. L. and Ball, J. J., J. Air Pollut.
      Contr. Ass., 23, 514 (1973).

193. Barth, C. A., Rusch, D. W. and Stuart, A. I., Radio Sci.,
      8, 379 (1973).

194. Perminova, D. N. and Shcherbov, D. P., Prom. Khim. Reakt.
      Osob. Chist. Veshchestv., 8, 181 (1967).

195. El-Ghamry, M. T., Frei, R. W. and Higgs, G. W., Anal. Chim.
      Acta, 47, 41 (1969).

196. Ryan, D. E. and Pal, B. K., Anal. Chim. Acta, 44, 385 (1969).

197. Babko, A. K., Terletskaya, A. V. and Dubovenko, L. I.,
      Zh. Anal. Khim., 23, 932 (1968).

198. Wheeler, G. L., Andrejack, J., Wiersma, J. H. and Lott,
      P. F., Anal. Chim. Acta, 46, 239 (1969).

199. Eichler, H., Z. Anal. Chem., 96, 22 (1934).

200. Powell, W. and Saylor, J., Anal. Chem., 25, 960 (1953).

201. Aguila, J. F., Talanta, 14, 1195 (1967).

202. White, C. E., in Fluorescence, Theory, Instrumentation,
      and Practice (G. G. Guilbault, Ed.), Dekker, New York,
      1967, p. 281.

203. Cherkesov, I. and Zhegalkena, V., Dokl. Akad. Nauk., SSSR,
      118, 309 (1958).

204. Kirkbright, G. F., West, T. S. and Woodward, C., Anal.
      Chem., 37, 137 (1965).

205. Gentry, C. and Scherrington, L., Analyst, 71, 432 (1946).

206. Nishikawa, Y., Hiraki, K., Morishige, K. and Shigematsu, T.,
      Japan Analyst, 16, 692 (1967).

207. de Albinati, J., Anales Assoc. Quim. Argentina, 53, 61
      (1965); Anal. Abstr., 5432 (1966).

208. White, C. E. and Lowe, C. S., Ind. Eng. Chem., Anal. Ed.,
      12, 229 (1940).

209. Will, F., Anal. Chem., 33, 1360 (1961).

210. Weissler, A. and White, C. E., Anal. Chem., 18, 530 (1946).

211. Davydov, A. and Devekki, A., Zavod. Lab., 10, 134 (1941).

212. White, C. E., McFarlane, H., Fogt, J. and Ruchs, R., Anal.
      Chem., 39, 367 (1967).

213. Holzbecher, Z., Chem. Listy, 47, 680, 1023 (1953).

214. Dale, A., Jones, P. and Radley, J., Inst. Rept., U.S. Dept.
      of Army Contract DA91-591-3309, 1965.

215. Dagnall, R. M., Smith, R. and West, T. S., Talanta, 13,
      609 (1966).

216. Goto, H., Sci. Rept. Tohoku Imp. Univ., Ser. 1, 29, 204
      (1940).

217. Goto, H., Sci. Rept. Tohoku Imp. Univ., Ser. 1, 29, 287
      (1940).

218. Goto, H., _Chem. Zb._, _1_, 1068 (1941).

219. Kirkbright, G. F., West, T. S. and Woodward, C., _Anal. Chim._
      Acta, _36_, 208 (1966).

220. Podberevskaya, N. K. and Sushkova, V., _Zavod. Lab._, _36_,
      1048 (1970).

221. Murata, A. and Ujaihara, T., _Bueneski Kagaku_, _10_, 497
      (1961); _Chem. Abstr._, _58_, 6180 (1964).

222. Marienko, J. and May, I., _Anal. Chem._, _40_, 1137 (1968).

223. Podchainova, V., Skonyakova, L. and Dvinyanimov, B., _Referat._
      Zh. Khim., _19GD_, 1968, Abstr. # 16G13.

224. Szelbellady, L. and Tamay, S., _Z. Anal. Chem._, _107_, 26
      (1936).

225. Radley, J., _Analyst_, _69_, 47 (1944); _Anal. Chem._, _21_, 1345
      (1949).

226. White, C. E., _J. Chem. Educ._, _28_, 369 (1951).

227. White, C. E., Weissler, A. and Busker, D., _Anal. Chem._, _19_,
      802 (1947).

228. White, C. E. and Hoffman, D. E., _Anal. Chem._, _29_, 1105
      (1957).

229. Parker, C. A. and Barnes, W. J., _Analyst_, _82_, 606 (1957).

230. Parker, C. A. and Barnes, W. J., _Analyst_, _85_, 828 (1960).

231. Szebelledy, L. and Gaal, F., _Z. Anal. Chem._, _98_, 255 (1934).

232. Marcantonatos, M., Gamba, G. and Monnier, D., _Helv. Chim._
      Acta, _52_, 538 (1969).

233. Tanbock, K., _Naturwiss._, _30_, 439 (1942).

234. Shcherbov, D. P. and Korzheva, R., _Tezisy dokladov sovesh-_
      chaniyapo lyuminestsenta, 1958, p. 65.

235. Kommenda, L., _Chem. Listy_, _47_, 531 (1953).

236. Holme, A., _Acta Chem. Scand._, _21_, 1679 (1967).

237. Neelakantam, K. and Row, L., _Proc. Ind. Acad. Sci._, _16a_,
      349 (1942).

238. Raju, N. and Rao, G., _Nature_, _174_, 400 (1954).

239. Babko, A. K. and Vasilevskaya, A., _Ukr. Khim. Zh._, _33_, 314
      (1967).

240. Rigin, V. and Melnichenko, N., Zavod. Lab., 33, 3 (1967).

241. Dancknortt, P. and Eisenbrand, J., Lumineszenzanalyze in filtrtierten ultravioletten Licht, Leipzig, 1956.

242. Korbl, Y. and Vydra, F., Chem. Listy, 51, 1457 (1957).

243. Wilkins, Y., Talanta, 4, 80 (1960).

244. White, C. E. and Lowe, C. S., Ind. Eng. Chem., Anal. Ed., 13, 809 (1941).

245. Fletcher, M. H., White, C. E. and Sheftel, M. S., Ind. Eng. Chem., Anal. Ed., 18, 179 (1946).

246. Holzbecher, Z., Coll. Czech. Chem. Commun., 20, 193 (1955).

247. Motojinia, K., Bull. Chem. Soc. Japan, 29, 75 (1956).

248. Sill, C. W. and Willis, C. P., Anal. Chem., 31, 598 (1959).

249. Klempeter, F. W. and Martin, A., Anal. Chem., 22, 828 (1950).

250. Laitinen, H. and Kivalo, P., Anal. Chem., 24, 1467 (1952).

251. Welford, G. and Harley, J., J. Amer. Ind. Hyg. Assoc. Quart., 13, 332 (1952).

252. Budesinsky, B. and West, T. S., Anal. Chim. Acta, 42, 455 (1968).

253. Naito, T., Nagano, H. and Yosui, T., Japan Analyst, 18, 1068 (1969).

254. Axelrod, H., Bonelli, J. and Lodge, J., Env. Sci. Technol., 5, 420 (1971).

255. Volman, V., Bull. Soc. Chim., 53, 385 (1933) (Ref. 8).

256. Borle, A. B. and Briggs, F., Anal. Chem., 40, 339 (1968).

257. Lewin, M., Wills, M. and Baron, D., J. Clin. Pathol., 22, 222 (1969).

258. Fingerhut, B., Poock, A. and Miller, H., Clin. Chem., 15, 870 (1969).

259. Classen, H. G., Marquardt, P. and Spath, M., Arzneimittel Forsch., 18, 211 (1968).

260. Uemura, T., Sci. Rept. Tohoku Imp. Univ., Ser. 4, 34, 31 (1968).

261. Miller, C. and Magee, R., J. Chem. Soc., 3183 (1951).

262. Bozhevol'nov, E. A., Federova, L., Krasavin, I. and Dziomko, V., Zh. Anal. Khim., 24, 531 (1969).

263. Louis, N. and Reber, A., Anal. Chem., 26, 936 (1954).

264. Eisenbrand, J., Pharma. Ztg., 75, 1003 (1930).

265. Patrovsky, V., Chem. Listy, 47, 676 (1953).

266. Bozhevol'nov, E. A., Chem. Abstr., 65, 7989 (1966).

267. Huu, C. Ti, Volkova, A. I. and Getman, T., Zh. Anal. Khim., 24, 688 (1969).

268. Hanker, J. S., Gamson, R. M. and Klapper, H., Anal. Chem., 29, 879 (1957).

269. Hanker, J. S., Gelberg, A. and Whitten, B., Anal. Chem., 30, 93 (1958).

270. Takanashi, S. and Tamura, Z., Chem. Pharm. Bull. Tokyo, 18, 1633 (1970).

271. Guilbault, G. G. and Kramer, D. N., Anal. Chem., 37, 918 (1965).

272. Guilbault, G. G. and Kramer, D. N., Anal. Chem., 37, 1395 (1965).

273. de Albinati, J., Anales Assoc. Quim. Argentina, 55, 61 (1967).

274. Temkina, V., Bozhevol'nov, E. A. and Dyatlova, N., Zh. Anal. Khim., 22, 1830 (1967).

275. Iritani, N., Miyahara, T. and Takahashi, I., Japan Analyst, 17, 1075 (1968).

276. Zeltser, L., Maksimyeheva, Z. and Talipov, S., Referat. Zh. Khim., 19GD, 1970, Abstr. # 1G85.

277. Konstantinov, A. V., Korobochkim, L. M. and Anastasina, G. V., Tr. Novoi Appl. Metod, 5, 167 (1967).

278. Bailey, B. W., Dagnall, R. N. and West, T. S., Talanta, 13, 1661 (1966).

279. Bozhevol'nov, E. A., Tr. VN11 Khim. Reakt., # 24, Goskhimy-dat, 1960.

280. Yamane, Y., Miyazaki, M. and Ohtawa, M., Japan Analyst, 18, 750 (1969).

281. Ritchie, K. and Harris, J., Anal. Chem., 41, 163 (1969).

282.   Pollard, F., McOmie, J. and Elberh, J., J. Chem. Soc.,
       446 (1951).

283.   Pollard, F., McOmie, J. and Elberh, J., J. Chem. Soc.,
       470 (1951).

284.   Shigematsu, T., Matsui, M. and Suimida, T., Bull. Inst.
       Chem. Res. Kyoto Univ., 46, 249 (1968).

285.   Shigematsu, T., Matsui, M. and Wake, R., Anal. Chim. Acta,
       46, 101 (1969).

286.   Fisher, R. P. and Winefordner, J. D., Anal. Chem., 43, 454
       (1971).

287.   Williams, D. E. and Guyon, J. C., Anal. Chem., 43, 139
       (1971).

288.   Melenteva, E., Poluektov, N. and Kononenko, L., Zh. Anal.
       Khim., 22, 187 (1967).

289.   Belcher, R., Perry, R. and Stephen, W. I., Analyst, 94, 26
       (1969).

290.   Willard, H. and Horton, C., Anal. Chem., 24, 862 (1952).

291.   Chaikin, S. W., Res. Ind., 5, # 3 (1953).

292.   Har, T. L. and West, T. S., Anal. Chem., 43, 136 (1971).

293.   Guyon, J. C., Jones, B. E. and Britton, D. A., Mikrochim.
       Acta, 1180 (1968).

294.   Tableaux des reactifs pour l'analyse minerale, report of
       the International Commission on New Reactions and
       Analytical Reagents, Paris, 1948.

295.   Fibk, D., Pivnichny, J. and Ohnesorge, W., Anal. Chem.,
       41, 833 (1969).

296.   Beck, G., Mikrochim. Acta, 47 (1939).

297.   Nazarenko, V. and Vinkovetskaya, S., Zh. Anal. Khim., 13,
       327 (1958).

298.   Ichihashi, M., Shigematsu, T. and Nishikawa, T., J. Chem.
       Soc. Japan, 78, 1139 (1957).

299.   Nishikawa, T., J. Chem. Soc. Japan, 79, 236 (1958).

300.   Sandell, E. B., Anal. Chem., 19, 63 (1947).

301.   Collat, J. and Rogers, L., Anal. Chem., 27, 961 (1955).

302. Bozhevol'nov, E. A., Lukin, A. and Gradinarskaya, M., USSR
      Pat. 116,838 (1958).

303. Lukin, A. and Bozhevol'nov, E. A., Zh. Akh., # 1 (1960).

304. Bozhevol'nov, E. A., Lukin, A., Yanishevskaya, V. and
      Kholod, E., USSR Pat. 119,287 (1958).

305. Bradaks, L., Feigl, F. and Hecht, F., Mikrochim. Acta, 269
      (1951).

306. Herzfeld, E., Z. Anal. Chem., 115, 131 (1939).

307. Bark, L. and Rixon, L., Anal. Chim. Acta, 45, 425 (1969).

308. Onishi, H., Anal. Chem., 27, 832 (1955).

309. Orighi, H. and Sandell, E. B., Anal. Chem. Acta, 13, 159
      (1955).

310. Shcherbov, D. P., Solovyan, I. and Drobachenko, A., Tezisy
      dokladov 6-go soveshchaniya lyuminestentu, Leningrad,
      1958.

311. Patrovsky, V., Chem. Listy, 48, 537 (1954).

312. Radley, J., Analyst, 68, 369 (1943).

313. Ladenbauer, I., Korkis, J. and Hecht, F., Mikrochim. Acta,
      1076 (1955).

314. Oshima, G., Japan Analyst, 7, 549 (1958).

315. Lukin, A. and Bozhevol'nov, E. A., J. Anal. Chem. USSR
      (English Transl.), 15, 45 (1960).

316. Bozhevol'nov, E. A., Lukin, A. and Gradinarskaya, M., Anal.
      Abstr., 7, 3164 (1960).

317. Raju, N. and Rao, G., Nature, 175, 167 (1955).

318. Tr. po Khim. i. Khim. Tekhnol., 1, 134 (1958).

319. Alford, W. C., Shapiro, L. and White, C. E., Anal. Chem.,
      23, 1149 (1951).

320. Brookes, A. and Townshend, A., Chem. Commun., 24, 1660
      (1968).

321. Oshima, G. and Nagasawa, K., Chem. Pharm. Bull. Tokyo,
      18, 687 (1970).

322. Ponomarenko, A., Markar'yan, N. and Komlev, A., Dokl.
      Akad. Nauk SSSR, 86, 115 (1952).

323. Shinagawa, N., Imai, H. and Sunabala, H., J. Chem. Soc. Japan, 77, 1479 (1956).

324. Bock, R. and Hochstein, K., Z. Anal. Chim., 138, 337 (1953).

325. Patrovsky, V., Chem. Listy, 47, 1338 (1953); Beck, G., Mikrochim. Acta, 287 (1937).

326. Bordea, A., Bull. Inst. Politech. Iasi, 13, 209 (1967).

327. Fink, D. and Ohnesorge, W., Anal. Chem., 41, 39 (1969).

328. Block, H., Paper Chromatography, 1955.

329. Pitts, A. and Ryan, D., Anal. Chim. Acta, 37, 460 (1967).

330. White, C. E., Fletcher, M. H. and Parks, J., Anal. Chem., 23, 478 (1951).

331. Michal, J., Chem. Listy, 50, 77 (1956).

332. Wallach, D. and Steck, T., Anal. Chem., 35, 1035 (1963).

333. Hill, J. B., Clin. Chem., 11, 122 (1965).

334. Schachter, D., J. Lab. Clin. Med., 54, 763 (1959).

335. Schachter, D., J. Lab. Clin. Med., 58, 495 (1961).

336. Hill, J. B., Ann. N. Y. Acad. Sci., 102, 1 (1962).

337. Klein, B. and Oklander, M., Clin. Chem., 13, 26 (1967).

338. Thiers, R. E., in Standard Methods of Clinical Chemistry (S. Meites, Ed.), Vol. 5, Academic Press, New York, 1965, p. 131.

339. Badrinas, A., Talanta, 10, 704 (1963).

340. Bozhevol'nov, E. A., Oesterr. Chem. Ztg., 66, 74 (1965); Chem. Abstr., 65, 7989 (1966).

341. Gusev, G., Lab. Delv, # 3, 157 (1968).

342. Dagnall, R. M., Smith, R. and West, T. S., Analyst, 92, 20 (1967).

343. White, C. E. and Cuttitta, F., Anal. Chem., 31, 2083 (1959).

344. Pal, B. K. and Ryan, D., Anal. Chim. Acta, 47, 35 (1969).

345. Kirkbright, G. F., West, T. S. and Woodward, C., Talanta, 13, 1637 (1966).

346. Kirkbright, G. F., West, T. S. and Woodward, C., _Talanta_, 13, 1645 (1966).

347. Titkov, Y., _Ukr. Khim. Zh._, 36, 613 (1970).

348. Andrushko, G., Maksimycheva, Z. and Talipov, S., _Referat. Zh. Chim._, 1969, Abstr. # 18G96.

349. Belman, S., _Anal. Chim. Acta_, 29, 120 (1965).

350. Sardesai, V. and Provido, H., _Mikrochem. J._, 14, 550 (1969).

351. Rubin, M. and Knott, L., _Clin. Chim. Acta_, 18, 409 (1967).

352. Sawicki, C., _Anal. Letters_, 4, 761 (1971).

353. Feigl, F., _Spot. Tests in Inorganic Analysis_ (tr. by R. E. Oesper), 5th Ed., Elsevier, Amsterdam, 1958.

354. Schenk, G., Dilloway, K. and Coulter, J., _Anal. Chem._, 41, 510 (1969).

355. Kautsky, H. and Hirsch, A., _Z. Anorg. Allgem. Chem._, 222, 126 (1935).

356. Franck, J. and Pringsheim, P., _J. Chem. Phys._, 11, 21 (1943).

357. Kautsky, H. and Miller, G., _Z. Naturforsch._, 2a, 167 (1947).

358. Tolmach, L. J., _Arch. Biochem. Biophys._, 33, 120 (1951).

359. Konstantinova-Shlesinger, M. A., _Tr. Fig. Inst. Akad. Nauk SSSR Fiz. Inst. imi P. M. Lebedevo_, 2, 7 (1942).

360. Konstantinova-Shlesinger, M. A. and Krasnova, V., _Zavod. Lab._, 6, 567 (1945).

361. Konstantinova-Shlesinger, M. A., _Zh. Fiz. Khim._, 9, 6 (1938).

362. Konstantinova-Shlesinger, M. A., _Chem. Abstr._, 30, 2521 (1936).

363. Watanabe, H. and Nakadoi, T., _J. Air Pollut. Control Assoc._, 16, 614 (1966).

364. Amos, D., _Anal. Chem._, 42, 842 (1970).

365. Land, D. B. and Edmonds, S., _Mikrochim. Acta_, 1013 (1966).

366. Kirkbright, G. F. and West, T. S., _Anal. Chem._, 43, 1434 (1971).

367. Lowry, O. H., Passonneau, J. V. and Schultz, S., _Anal. Biochem._, 19, 300 (1967).

368. Veening, H. and Brandt, W., _Anal. Chem._, _32_, 1426 (1960).

369. Axelrod, H., Cary, J., Bonelli, J. and Lodge, J., _Anal. Chem._, _41_, 1856 (1969).

370. Morgan, E., Vlasov, N. and Tyutin, V., _Referat. Zh. Khim. 19GD_, 1969, Abstr. # 23G162.

371. Komlev, O. and Zinchuk, V., _Referat. Zh. Khim. 19GD_, 1968, Abstr. # 4G66.

372. Ivankova, A. I. and Shcherbov, D. P., _Referat. Zh. Khim._, 1968, Abstr. # 18G104.

373. Filer, T. D., _Anal. Chem._, _43_, 725 (1971).

374. Nishikawa, Y., Hiraki, K. and Shigematsu, T., _J. Chem. Soc. Japan_, _90_, 483 (1969).

375. Nazarenko, V. and Antonovich, V., _Zh. Anal. Khim._, _24_, 358 (1969).

376. Shcherbov, D. P. and Nikolaeva, V., _Prom. Khim. Reakt. Osob. Chist. Veshchestv._, 186 (1967).

377. Cousins, E. B., _Austral. J. Expt. Biol. Med. Sci._, _38_, 11 (1960).

378. Watkinson, J. H., _Anal. Chem._, _32_, 981 (1960).

379. Parker, C. A. and Harvey, L. G., _Analyst_, _86_, 54 (1961).

380. Parker, C. A. and Harvey, L. G., _Analyst_, _87_, 558 (1962).

381. Allaway, W. H. and Cary, E. E., _Anal. Chem._, _36_, 1359 (1964).

382. Lott, P. F., Cukor, P., Moriber, G. and Solga, J., _Anal. Chem._, _35_, 1159 (1963).

383. Clarke, W. E., _Analyst_, _95_, 65 (1970).

384. Wilkie, J. B. and Young, M., _J. Agr. Food Chem._, _18_, 946 (1970).

385. Olson, O., _J. Assoc. Off. Anal. Chem._, _52_, 627 (1969).

386. Kasiura, K., _Chemia Anal._, _14_, 1325 (1969).

387. Kononenko, L. I., Melenteva, E., Vitkin, R. and Polvektov, N., _Prom. Khim. Reakt. Osob. Chist. Veshchestv._, 223 (1967).

388. Coyle, C. F. and White, C. E., _Anal. Chem._, _29_, 1486 (1957).

389. Pal, B. K. and Ryan, D., Anal. Chim. Acta, 48, 227 (1969).

390. Tishchenko, M., Kononenko, L. I. and Poluektov, N., Prom. Khim. Reakt. Osob. Chist. Veshchestv., 231 (1967).

391. Dagnall, R. M., Smith, R. and West, T. S., Analyst, 92, 358 (1967).

392. Butter, E., Kolowos, I. and Holzapfel, H., Talanta, 15, 901 (1968).

393. Kononenko, L. I., Mishchenko, S. and Poluektov, N., Zh. Anal. Khim., 21, 1392 (1966).

394. Bottei, R. S. and D'Alessio, A., Anal. Chim. Acta, 37, 405 (1967).

395. Milkey, R. and Fletcher, M., J. Amer. Chem. Soc., 79, 5425 (1957).

396. Babko, A., Hzeu, C., Volkova, A. and Getman, T., Ukr. Khim. Zh., 35, 292 (1969).

397. Feigl, F., Gentil, V. and Goldstein, D., Anal. Chim. Acta, 9, 393 (1953).

398. Konstantinova-Shlesinger, M. A., Fluorometric Analysis (N. Kaner, Transl.), Davey, New York, 1965, p. 159.

399. Tomic, E. and Hecht, F., Mikrochim. Acta, 896 (1955).

400. Rao, V. and Rao, G., Z. Anal. Chim., 161, 406 (1958).

401. Bottei, R. S. and Trusk, A., Anal. Chim. Acta, 41, 374 (1968).

402. Murata, A. and Yamaguchi, F., J. Chem. Soc. Japan, 77, 1259 (1956).

403. Kirillov, A., Lauer, R. and Polvektov, N., Zh. Anal. Khim., 22, 1333 (1967).

404. Ichihashi, M., Shigematsu, T. and Nishikawa, T., J. Chem. Soc. Japan, 77, 1474 (1956).

405. White, C. E. and Neustadt, M., Ind. Eng. Chem., Anal. Ed., 15, 599 (1943).

406. Merritt, L., Ind. Eng. Chem., Anal. Ed., 16, 758 (1944).

407. Yoe, J. H. and Jones, A. L., Ind. Eng. Chem., Anal. Ed., 16, 111 (1944).

408. Vosburgh, W. C. and Cooper, G. R., J. Amer. Chem. Soc., 63, 437 (1941).

409. Harvey, A. E. and Manning, D. L., *J. Amer. Chem. Soc.*, 72, 4488 (1950).

410. Harvey, A. E. and Manning, D. L., *J. Amer. Chem. Soc.*, 74, 4744 (1952).

411. Bjerrum, J., *Metal Ammine Formation in Aqueous Solution*, P. Haase and Son, Copenhagen, 1941.

412. Job, P., *Ann. Chim. (Paris)*, 9, 113 (1928).

413. Guttman, D. and Gadzala, A. E., *J. Pharm. Sci.*, 54, 742 (1965).

414. Steiner, R. F., Roth, J. and Robbins, J., *J. Biol. Chem.*, 241, 560 (1966).

415. Naik, D. V., Paul, W. L. and Schulman, S. G., *Anal. Chem.*, 47, 267 (1975).

416. Naik, D. V., Paul, W. L. and Schulman, S. G., *J. Pharm. Sci.*, In Press.

417. Scatchard, G., *Ann. Y. Y. Acad. Sci.*, 51, 660 (1949).

418. Edsall, J. T., Felsenfeld, C., Goodman, D. S. and Gurd, F. R. N., *J. Amer. Chem. Soc.*, 76, 3054 (1954).

419. Rees, V. H., Fildes, J. F. and Laurence, D. J. R., *J. Clin. Pathol.*, 7, 337 (1954).

420. Davison, C., in *Fundamentals of Drug Metabolism and Drug Disposition*, B. N. La Du, H. G. Mandel and E. L. Way, Eds., Williams and Wilkins, Baltimore, 1971, Chapt. 4.

421. Meyer, M. C. and Guttman, D. E., *J. Pharm. Sci.*, 57, 895 (1968).

422. Potter, G. D. and Guy, J. L., *Proc. Soc. Exp. Biol. Med.*, 116, 658 (1964).

423. F. E. Hahn, Ed., *Progress in Molecular and Subcellular Biology, Vol. 2, Complexes of Biologically Active Substances with Nucleic Acids and Their Modes of Action*, Springer-Verlag, New York, 1971.

424. Blake, A. and Peacocke, A. R., *Biopolymers*, 6, 1225 (1968).

425. Van Duuren, B. L., *Chem. Revs.*, 63, 325 (1963).

426. Peacocke, A. R. and Skerrett, J. N. H., *Trans. Faraday Soc.*, 52, 261 (1956).

427. R. C. Nairn, Ed., *Fluorescent Protein Tracing*, Livingstone, Edinburgh, 1962.

428. Weber, G., _Biochem. J._, 51, 155 (1952).

429. Stryer, L. and Haugland, R. P., _Proc. Nat'l. Acad. Sci._
      _U.S._, 58, 67 (1967).

430. Levin, A., Killander, D., Klein, E., Nordenskjöld, B. and
      Inoue, M., _Ann. N. Y. Acad. Sci._, 177, 481 (1971).

INDEX